COMPETITION, LTD.:

The Marketing of Gasoline

COMPETITION, LTD.:

The Marketing of Gasoline

Fred C. Allvine & James M. Patterson

INDIANA UNIVERSITY PRESS

Bloomington / London

Published in Canada by Fitzhenry & Whiteside Limited, Don Mills, Ontario

Library of Congress catalog card number: 70-180491

ISBN: 0-253-31390-2

Manufactured in the United States of America

Contents

Tables

Figures

Plates

(Following page 61)

Preface

The limitations of classical economical models for studying market structure and institutional change led the authors to look for a different approach. This book has its intellectual origins in an insightful study by Joseph Palamountain entitled *The Politics of Distribution.*[1] Palamountain's concept of intertype competition is ideally suited for the study of competition in the marketing of gasoline to the public.

An interest in trading stamps and games of chance as promotional instruments exposed the authors to the gasoline industry with its complex array of problems. Curiosity about certain analomies that existed in the gasoline industry led the authors to do some preliminary research which finally resulted in the two year study supporting this book. Once we began the study it became clear that the marketplace was providing relatively little direction to competition; instead it was government policy that seemed to be the more important determiner of what was occurring. Furthermore, it soon became apparent that vital intertype competition was being thwarted, leading to great inefficiencies in the marketing of gasoline.

In search of support for the study, the Society of Independent Gasoline Marketers was approached on the chance that they might fund an independent study of competition in the marketing of gasoline. They were willing to make a generous grant to Northwestern University to support the inquiry. In addition, over forty of the key figures in the independent sector of the gasoline industry were willing to explain in detail their marketing strategies and marketing problems. Interviews were also held with some major oil company marketing executives, but

1. Joseph C. Palamountain, *The Politics of Distribution* (Cambridge: Harvard University Press, 1955).

as one might expect, this was more difficult given the prevailing critical mood toward business—especially big business.

While we are indebted to many many people several deserve our special recognition for their help and counsel. A special thanks goes to Ron Peterson, chairman of the board of Martin Oil Service Company, for his inspiration and frequent help. Howard Teak, SIGMA's executive director, opened many doors. Carl Greer, Newell Baker, and Louis Kincannon read various parts of the study for factual and institutional errors. Dean Alfred E. Kahn of Cornell, a distinguished student of the petroleum industry, reviewed the manuscript and made numerous helpful suggestions. Needless to say, any remaining flaws are of our doing and not theirs. We are also grateful to the Graduate School of Management of Northwestern University and to the Graduate School of Business at Indiana University for providing much needed institutional support. Very much appreciated are the efforts of Mrs. Marion Davis and Mrs. Edith Bass at Northwestern University who typed countless versions of the manuscript and those of our editor, Sheila Steinberg, of the IU Press.

COMPETITION, LTD.:

The Marketing of Gasoline

1 / *An Introduction to Gasoline Marketing and Its Problems*

THE OIL BUSINESS is no stranger to the spotlight of critical examination. The public's interest in this major source of energy is to be expected. In recent times, the impact of motor fuel emissions and oil spills on the natural environment has received considerable attention. The importance of petroleum self-sufficiency to national security and international relations is obvious. The influence of vested oil interests on domestic politics has provided grounds for scores of scathing studies. The fact that oil is a limited natural resource, consumed in use, raises important issues of conservation. The rash of mergers among regional oil companies, the hassle over the depletion allowance, and the cabinet-level review of the mandatory oil import program raise important economic questions. All these dimensions of the oil business—economic, political, national security, and conservation—endow the industry with public interest.

In addition, the oil business is big business. Seven of the top 20 of *Fortune*'s 500 largest U.S. corporations are oil companies. Oil companies have figured importantly in the history of antitrust laws and in the evolution of our national commitment to maintain competition as the principal regulator of economic activity. The great Standard Oil Trust was influential in bringing about the passage of the Sherman Antitrust Act in 1890, and the subsequent break-up of the trust in 1911 was the high point of Roosevelt's trust-busting activity. The oil business thus raises additional important questions of economic power and its regulation.

As one would expect, the oil industry has been studied and restudied. There are important books which deal with the economic, political, conservation, and national-security issues raised by the industry. Less well studied has been the marketing of gasoline, the most important product produced from petroleum. The Federal Trade Commission released a report in 1967 on the *Anticompetitive Practices in the Marketing of Gasoline*[1] and previously had conducted an extensive industry conference on gasoline marketing.[2] Also the Committee on Small Business in the House and the Subcommittee on Antitrust and Monopoly in the Senate have periodically inquired into the marketing of gasoline and have critically examined marketing practices of the industry. But, by and large, the marketing aspect of the industry has not been examined as carefully as the other aspects of the petroleum business. Nor are the interrelationships of marketing with these other aspects clearly understood. This is a serious omission, since attempts to look at marketing out of its industry context can often be misleading. In part this study seeks to overcome this omission.

Along with production, refining, and transportation, marketing is an important part of the industry. It is also important to the consuming public. The average automobile is driven 9,800 miles per year and consumes 731 gallons of gasoline. Out of every dollar spent by individual consumers in 1970, 3.7¢ were spent on gasoline and oil. This was 8.7¢ of every dollar spent on nondurables. Furthermore, the dollar investment in 220,000 service stations and the attendant employment are substantial.[3] Thus, the way gasoline is marketed is important, both to the industry and to society at large. If it is inefficient or unresponsive, a vast amount of economic resources and the well-being of many people are at stake.

This study was undertaken to explore whether competition is working as an effective regulator of this industry. The preliminary research indicated that many deep-seated problems affect the way in which gasoline is being sold to the public. In contrast to what is occurring in several other industries, mass-merchandising techniques seem to be shunned in the marketing of gasoline. During the later 1940's and 1950's mass merchandising was adopted in the grocery business as supermarket outlets deeply penetrated this industry. Frequently one large and efficient supermarket would replace five or ten smaller neighborhood stores. In the process, product assortment increased, physical facilities improved, and prices decreased. Similarly, during the 1960's, discount stores grew rapidly by adopting the mass-merchandising techniques in the marketing of general merchandise. However, in the petroleum industry the mass merchandising of gasoline on a high-volume, low-price basis has made relatively little headway. Those who seemed to be trying to lead the industry in this direction have been confronted with many obstacles. In certain parts of the country, the trend has actually been away from the mass merchandising of gasoline. The general prevailing technique for selling gasoline to the public continues to be through large numbers of highly convenient single-brand gasoline outlets. This anomaly in the way gasoline is marketed was a major motivating factor in undertaking this study of the vitality of competition in the marketing of gasoline.

A central focus of this study is on the major forces of competition in the marketing of gasoline. The study explores what role intertype competition—the competition between major oil companies (e.g., Texaco, Shell) and private branders (e.g., Hudson, Martin, Merit)—plays in this industry. Are the private branders as important a force for preserving competition as suggested by the Federal Trade Commission report? [4] If so, is damage to intertype competition as a result of business conduct, or the structure of the industry, a problem with which to be concerned? Our assumption in this study is that intertype competition critically affects the vitality of competition in any line of business. This proposition as it relates to the gasoline industry needs to be examined.

This study also had its origins in collateral questions about developments in the gasoline-petroleum industry. Clearly, important structural changes were being wrought by the merger of major firms such as Union with Pure, Atlantic with Richfield and then with Sinclair, and Sun with Sunray DX. The Justice Department approved them on the ground that they were market-extension mergers and thus would not harm competition. But is this true? Sinclair and Sunray DX were both principal suppliers of unbranded gasoline to private-brand marketers. After merging with Arco and Sun they have not been as eager to sell to private branders. Perhaps the mergers have harmed competition after all by contributing to a reduction in sources of supply of gasoline to the private-brand marketers. This proposition needs to be explored.

Not only is there a question concerning the impact on intertype competition of mergers of regional oil companies, but also one of the effect of direct purchase of private branders by the majors. Again these purchases tended to go unchallenged on the ground that many of them were market-extension mergers where supposedly there would be little or no harm to competition. Frequently in markets like Chicago, or large areas like the West Coast, the purchase of one private brander by a major oil company was followed by still another and another. Often the number of sellers in each market remained approximately the same. The basic change was that the small local marketers were replaced by major oil companies new to the markets. While in some cases the identity and price-discounted method of operation of the private brander were maintained, more often the major oil companies converted the purchased independent to their brand-name and non-price method of marketing. The consequences of the buy-outs of private branders need to be further explored.

Are the periodic price wars in certain areas evidence of vigorous competition or something else? Are majors using deep price cuts to discipline and otherwise regulate or destroy price marketers? Price wars, by definition, are abnormal. Do they also destroy competition? Is the motorist better off or worse off as a result of price wars? In the short run he is clearly better off since he pays less for gasoline, but what about the long run? If the smaller, financially weaker, though not necessarily less efficient private branders are driven out of business or otherwise taught to be less price aggressive by price wars, what will happen to prices over the long haul? A lot depends on how easy it is for new price marketers to enter the business after others have departed and prices have been increased. Furthermore, to what extent do contracting sources of supply for the private branders, rising costs of new locations, and the ever-present prospect of renewed price wars serve to deter new entry? Clearly the impact of price wars on long-run consumer interest needs to be explored.

As a related question one must ask if competition in gasoline is fair? While this is hard to answer, it is still a legitimate question. For example, is it fair for major oil companies to reduce their prices to dealers in a specific zone surrounding a price marketer while holding their prices constant in less competitive market areas? This practice permits the huge geographically and functionally diversified firm to focus its superior financial strength on specific, less powerful competitors. Can we expect competition to flourish if this practice is permitted? On the other hand, if it were restricted, would we merely be preserving competitors at the expense of price competition? The answer is not clear.

In this industry one is also troubled by the apparent waste in retail distribution. Most service stations are fantastically underutilized. The intensive distribution strategies followed by each of the major firms in this industry lead collectively to substantial excess capacity. Anyone familiar with the problems in gasoline marketing will admit that there are far too many stations. When the available gasoline business is divided among them, few stations can reach efficient levels of manpower and facility utilization. The result is low labor productivity and low profitability. Evidence of this inefficiency is given by the fact that dealer turnover among major-brand dealers often exceeds 25 percent each year. This despite the fact that dealer gross margins are relatively high given the simple retail function that is performed. Normally, if market forces were allowed to work, this overcapacity would quickly be eliminated. Those stations unable to earn a reasonable return would be closed. Why this costly unutilized capacity continues to hang over the market needs to be explored.

The effects of vertical integration on the marketing methods of the major oil companies must also be considered. In the petroleum industry practically all of the major-brand marketers are vertically integrated. They are involved in the production of crude oil, the operation of pipelines, the running of refineries, the distribution of refined products, and the marketing of gasoline over broad geographic areas. By contrast most independents operate primarily at one or two levels of integration with private branders operating only at the marketing level.

There are two aspects of vertical integration which need to be examined. First, to what extent does vertical integration color the marketing style of the major oil companies? Without vertical integration does the present marketing style of the majors make any sense? This is a marketing style based upon selling branded gasoline through large numbers of highly convenient, controlled, single-brand outlets, with the primary emphasis on nonprice methods of marketing, and where price discounting is normally avoided.

A second concern about vertical integration is to what extent it poses a serious threat to the survival of the nonintegrated or partially integrated firm seeking to compete at only the refinery and marketing levels. Given the structure of the industry as it exists with incentive and opportunity to capture disproportionately large profits in the crude-oil department, the integrated companies are put into a position where it is highly probable that they will subsidize their refining and marketing activities—the downstream operations. In some cases it may not be undesirable for profits from the nonretail level of an integrated firm to be used to subsidize the retail level of the business. For example, one should be delighted when an otherwise sensible firm takes good money out of the bank and distributes it to its retail customers by means of retail price cuts. Nor should it be too upsetting if some of its competitors fall by the way as a result of not being able to meet these low prices. If the subsidizing firm later decides to stop the subsidy and raise prices, it is anticipated that either the old competitor or entirely new ones will reenter the market and check the price increase. As long as entry is easy, the subsidy may be unfair, but not antisocial. However, should entry not be easy, or should the integrated firm make it clear that it will drive out new entrants in the same fashion, the practice takes on an entirely different cast. The effects of subsidizing downstream operations in the oil industry on the long-run vitality of competition in the marketing of gasoline deserve careful appraisal.

One of the consequences of vertical integration is that at the very least it insulates operations at any one level from those competitive forces which would otherwise guide investment and behavior. The best way to determine retail margins or distributive investment or even the character of the retail offer is through the interplay of competition. But vertical integration diffuses and blunts these corrective forces. It is the profitability of the entire vertical system that governs behavior, not the profitability of operations at any one level. Most gasoline markets contain enough different competitors to effectively regulate marketing margins and practices. Were it not for vertical integration the retailing of gasoline would probably be reasonably competitive. The proposition that many marketing aberrations would disappear if vertical integration were not present needs to be assessed.

Retail gasoline competitors, whether they be major-brand stations, private branders, or the in-between semimajors, live in a delicate balance with one another. However, the ability of private branders to compete on an equal footing seems to be affected by the special privileges that some receive and others do not. Those vertically integrated firms producing crude oil enjoy special tax privileges and refiners receive valuable import tickets permitting them to purchase low-cost foreign crude oil. In a sense, these benefits represent an indirect government subsidy to the integrated firm which is not available to the specialized retailer and which may distort the normal forces of competition at work in the market place. Furthermore, given the beneficial tax laws affecting crude-oil profits, the integrated firm has strong incentive to arrange interlevel transfer prices so that profits from the total system accrue primarily at the crude-oil production level. Those marketing distortions created by special privileges received by some and not by others must be examined.

Another important question in assessing industry performance is whether the competitive practices of the major oil companies that dominate the industry have tended to preserve or to undermine consumer sovereignty—consumer choice between reasonable purchase alternatives. The prevailing market strategy of the major oil companies has been one of trying to build brand preference for something that is close to being a homogeneous product in terms of its physical properties. The differentiation achieved has been created by managing the quality and character of the retail setting and by the use of audio and visual symbols. Over the years this method of marketing has become more committed to costly nonprice techniques—credit cards, stamps, premium offers, fancier stations, and the like. For that portion of the population genuinely desiring such an approach this method of marketing is fine. However, are those gasoline customers who prefer something different given reasonable purchase alternatives? On the other hand, are alternative merchandising approaches forcefully resisted by the dominant members of the industry? Do they employ unfair competitive practices to thwart the effort of companies that choose to streamline their operation to achieve volume sales and mass-merchandise efficiencies by selling for less? Are new purchase propositions such as self-service and convenience outlet tie-ins permitted a fair test of the market place? Are retail variety and enlightened choice frustrated by the present business conduct and the structure of the industry? If so, government intervention may be necessary to preserve consumer sovereignty and to restore to the market place its role of regulating competitive practices.

In the chapters that follow the issues raised in the Introduction are explored. The questions treated in the remainder of the book are outlined below.

1. What is the fundamental marketing strategy of the major oil companies? What are the different elements in the strategy and how are they related?
2. Why have the private branders chosen to market gasoline as

they do? How is one to explain their emergence and evaluate their prospects? What is the economic logic of the private branders' operations?

3. What competitive role does the private brander play in the retailing of gasoline? Why should anyone be interested if this type of operation survives? Would consumer interest be preserved if competition were primarily between a dozen large integrated firms?
4. What causes price wars? Are they in the public interest and what is their effect on competition?
5. Are the price-protection programs of the major oil companies in the public interest? Would eliminating the practice of granting price protection make competition more vigorous? Is zone price protection as it is being employed a legal competitive tool?
6. Is the buy-up of private branders by the majors destructive of intertype competition? Does it make any difference how the majors subsequently operate the stations they have purchased?
7. How have recent mergers of major oil companies affected the private branders? Can the private-brander's interest in the independence of the merged companies be preserved without disallowing the merger?
8. Are problems in the gasoline industry related to integration itself? Do the integrated firms have an unfair advantage which gives them an edge in retail competition?
9. Do the major oil companies in fact subsidize their retail operations? If this is so, is this bad for competition and contrary to the public interest?
10. To what extent is availability of supply important to the continued survival of the independent? Why do some major oil companies supply price marketers, while others do not? Should the price marketer be guaranteed an adequate supply of unbranded gasoline at an economical price? If so, how?

In the following chapters the forces that have shaped, and are currently changing, the ways in which gasoline is marketed will be presented. By no means is this a frictionless industry where new ideas and more efficient methods of marketing gasoline are readily adopted and old approaches are laid quietly to rest. It is hoped that this book will present new insights into complex and deep-seated problems in this industry, and that the book will ultimately contribute to improvement in the competitive performance of the gasoline industry.

2 / *Marketing Style of the Major Oil Companies*

When a driver stops at a major-brand service station to purchase gasoline his experience is likely to be as follows. The station will probably be just one of the many conveniently located service stations selling a single brand of nationally advertised gasoline. He will not shop for price, since he knows that most major-brand gasolines are normally sold for about the same price. While the gasoline tank is being filled, the attendant will wash the car windows and check the oil and water on request. The driver then uses a major oil company credit card to pay for his purchase, receiving in turn a handful of paper—a credit card sales slip, 70 trading stamps, and a game card. While all this is going on he may have noticed that minor repair work is being performed on cars in two of the three service bays.

It probably seems quite natural to the driver for gasoline to be sold through service stations in this way. However, in reality many dimensions of this method of selling gasoline are neither natural nor preordained. Often the components are the results of calculated decisions by major oil companies to gain some measure of control over the final demand for their most important and valuable product—gasoline.

One would be misled, however, if he thought that the major oil companies made their decision independently of the structure of the industry within which they compete. In this industry, as in others, the industry setting suggests an appropriate marketing style which is generally adopted by the majority of successful firms. Thus the emphasis on product differentiation, credit cards, continuity promotions, and exclusive outlets reflects the market structure of the petroleum industry as well as the way in which competitors have decided that it is to their advantage to operate.

STRUCTURAL FEATURES OF THE GASOLINE INDUSTRY

ONE OF THE IMPORTANT STRUCTURAL FEATURES of the industry that dictates the nature of retail competition is the domination of the industry by fifteen to twenty huge oil companies. These companies refine approximately 85 percent of the domestic gasoline and sell approximately 80 percent of all gasoline purchased by drivers through their branded stations. The organization, nature of operation, and problems of these oil companies are very similar and as a result their marketing practices closely parallel one another. The way in which the majors have chosen to operate is clearly the dominant style.

When a market is dominated by a few large firms—the textbook term is oligopoly—price competition is severely constrained by the fact that each firm is too large a factor in the market to be ignored by the others were it to cut its price in order to increase its business. Each of the others would have to meet the price cut. If the total amount of the product that is bought at the lower price is not substantially greater—which is the case with gasoline—all firms, including the initial price cutter, will be worse off. Consequently, given this kind of structure, price cuts are generally eschewed and competitive rivalry for a share of the market assumes a variety of nonprice forms.

A second important structural feature of the industry is that all the major firms are vertically integrated. These firms are not only engaged in distributing and marketing gasoline, but also in finding and producing crude oil from which gasoline is made, in running tremendous refineries where crude oil is made into gasoline and other oil products, and finally in operating complex networks of pipelines for controlled distribution of crude oil and refined products. When firms are vertically integrated, one division of a company sells to another rather than purchasing its requirements on the open market. Transfer prices and hence gross margins at each level are set by managerial action rather than by market forces. The result is that the profit at any particular level can be manipulated as the logic of the integrated operation dictates. This fact heavily influences the marketing strategy of the integrated firms and raises havoc for the nonintegrated refiners and the independent marketers. The squeezing of margins at refining and at marketing, as well as the subsidization of low returns at one level by high returns at another, are part and parcel of the strategy of the majors and reflect the logic of vertical integration.

A third structural feature, and possibly the most important one for understanding much of what is happening in this industry, is the role of crude oil. Those companies producing a larger portion of the crude oil than is needed by their refineries have an important competitive edge over those buying a large portion of their crude-oil requirements from others. The advantage of those companies with strong crude-oil positions results from the fixed and high price of crude oil. Their competitors who are poor in crude oil have to pay an artificially high price for raw material, which reduces their ability to compete. Many of the marketing practices, or as one major oil company executive says "the marketing overkill," are associated with the drive to cash crude-oil profit. The frequent price wars that plague the industry in certain areas of the country are in part financed by crude-oil profits. Finally, the industry is becoming more concentrated as those poor in crude oil are compelled to merge with those more fortunate.

The Role of Price

While the major oil companies emphasize nonprice methods of competition, this does not mean that prices can be ignored. Quite the contrary! Workable nonprice competition requires that major oil companies and their dealers refrain from lowering prices as an active instrument of competition. As a result a great deal of the majors' effort is directed toward insuring that major-brand competitors do not use lower prices to gain a competitive edge. Unless the majors generally agree to compete on a fundamentally nonprice basis, a market is likely to remain in a continuous state of price chaos, with tremendous marketing losses for all involved. Prices must be relatively high and stabilized for the majors to implement their nonprice method of marketing gasoline.

Categories of Majors

To understand how prices are established it is helpful to divide the major oil companies into three categories: market leaders, important majors, and minor majors. These three categories of majors interact with one another in establishing market prices.

In a specific market, the market leaders generally are those competitors who have relatively large numbers of stations, good market acceptance, and a large share of the market. Because of their market position, they are particularly interested in a stable market at a relatively high price. As a result of their strength in a market, the market leaders have principal responsibility for establishing and maintaining what is referred to as the market reference price. This is the price which major-company dealers are generally expected to follow if there is to be price stability in a market and profitable retailing operations. Typically the market leaders have the largest number of their stations selling at or above the market reference price.

For example, the market leaders in Washington, D.C. are Esso, American, and Shell (see Table 2-1). They have the largest number of stations (with the exception of Texaco), the smallest proportion of stations below reference and posting price, and the largest proportion of stations above reference price. Clearly these three brands are making the reference price of 35.9¢ a gallon for regular gasoline a relevant price in the market. Many of the same conditions were observed for the market leaders in San Francisco (see Table 2-2) and St. Louis (see Table 2-3).

An analysis of market leaders, however, suggests that this category of major oil companies might, in some cases, be divided into two subgroups—the reference marketer and the aggressive leader. The reference marketer is more inclined to turn his back on off-pattern pricing and, therefore, tends to have only a relatively small portion of his stations off pattern and posting price. This can be seen from Table 2-4. On the other hand, the aggressive leader has a much greater proportion of his stations below reference and posting price. The aggressive leaders acts as a policing agent who comes down hard on those who significantly get out of line. Certain majors such as Shell seem to be in a better position to act as an aggressive leader. Shell has closed many of its smaller stations and has opened larger centrally located stations where periodic price cuts help stations to build needed gasoline sales volume. Also Shell has a widely dispersed operation so that should its aggressiveness cause a price war in some of its markets, it would not have a major effect on Shell's overall operation. On the other hand, Union, Chevron, and, to a lesser extent, American are concentrated in

TABLE 2-1

Price Survey of Stations Located Throughout Washington, D.C.—Fall 1969

	Regular 29.9¢	30.9¢		31.9¢	32.9¢	33.9¢		34.9¢ (−1¢)		35.9¢ (Major Reference Price)		36.9¢ (+1¢)		37.9¢	Stations by Brands		Pricing Below Refer-ence	Posting Prices on Signs
Brand	Premium 33.9¢	33.9¢	34.9¢	35.9¢	36.9¢	37.9¢	38.9¢	38.9¢	39.9¢	39.9¢	40.9¢	39.9¢	40.9¢	41.9¢	No.	%		
Esso					2	2	7	7	3	61	9	1	50	2	144	20.7	14.6%	8.3%
American							1		13	5	53		17	1	90	12.9	15.6	4.4
Shell					1	3		5	1	50	2	3	7	1	73	10.5	13.7	2.7
Sunoco								8	2	47	1	3			61	8.8	16.4	49.2
Gulf								14		36	3	3	1	2	59	8.5	23.7	27.1
Texaco					2		3	12	7	33	3	5	1		66	9.5	36.4	15.1
Mobil			1		1	4	2	18	1	19	2	1	1		50	7.2	54.0	34.0
Sinclair			2		1	1	3	8	2	10	2	2	1		32	4.6	53.1	3.1
Citgo					1	3		8		14			2		28	4.0	42.9	7.1
Atlantic					2		1	16	3	4					26	3.7	84.6	15.4
Phillips Chevron					1	1		5		7					14	2.0	50.0	14.3
Major No.			3		11	14	17	101	32	286	75	18	80	6	643	92.3	27.7%	15.5%
Subtotal %			(0.5%)		(1.7)	(2.2)	(2.6)	(15.7)	(5.0)	(44.5)	(11.7)	(2.8)	(12.4)	(.9)	(100.0)			
Hess			9												9	1.3		100.0
Scot	1		3	10	1										15	2.2		80.0
Crown		6													6	.9		66.7
Miscellaneous																		
Independents	7	4	4	7	1			1							24	3.4		79.2
Subtotal No.	8	10	16	17	2			1							54	7.8		81.5%
Price																		
Marketers %	(14.8%)	(18.5)	(29.6)	(31.5)	(3.7)			(1.9)							(100.0)			
Total	8	10	19	17	13	14	17	102	32	286	75	18	80	6	697	100.0		

Source: Survey conducted by authors for this study.

TABLE 2-2

Price Survey of Stations Located Throughout San Francisco and the West Bay Area—Fall 1969

	Regular	27.9¢	28.9¢	29.9¢	30.9¢	31.9¢		32.9¢		33.9¢		34.9¢ (−1¢)		35.9¢ (Major Reference Price)		36.9¢ (+1¢)		37.9¢	Stations by Brands		Pricing Below Refer- ence	Posting Prices on Signs
Brand	Premium	30.9¢	31.9¢	32.9¢	33.9¢	34.9¢	35.9¢	35.9¢	36.9¢	36.9¢	37.9¢	37.9¢	38.9¢	38.9¢	39.9¢	39.9¢	40.9¢	40.9¢	No.	%		
Chevron												1	5		33	21	11	15	86	20.2	7.0%	5.8%
Shell								1	1	4		11		10	1	17	2	8	55	12.9	30.9	32.7
Union														1	6	18	11	11	47	11.1		4.3
Phillips								1			1		5	6	2	19	3	11	48	11.3	14.6	10.4
Richfield												1	7	3	10	16	2	2	41	9.6	19.5	26.8
Mobil									1		1	2	2	5	2	13	3	4	33	7.8	18.1	30.3
Texaco					1		1	1	1	2	2	4	5	4	2	8	2	2	35	8.2	48.6	37.1
Gulf							2		2		9		7		1		1		22	5.2	91.0	68.2
Esso					3				5	1	1		2		2		1	1	16	3.8	87.4	50.0
American						1													1	.2	100.0	100.0
Subtotal No.					4	1	3	3	10	7	14	19	33	29	59	112	36	54	384	90.3	24.5	22.9
Majors %					(1.0)	(0.3)	(0.8)	(0.8)	(2.6)	(1.8)	(3.6)	(4.9)	(8.6)	(7.5)	(15.4)	(29.2)	(9.4)	(14.1)	(100.0)			
Major Secondaries						1	8	1	2		7	2	1						22	5.2	100.0	90.9
Independents		3		3	3	2	3			1	1	1	2						19	4.5	100.0	94.8
Subtotal No.		3		3	3	3	11	1	2	1	8	3	3						41	9.7		92.7
Price Marketers %		(7.3%)		(7.3)	(7.3)	(7.3)	(26.8)	(2.4)	(4.9)	(2.4)	(19.5)	(7.3)	(7.3)						(99.8)			
Total		3		3	7	4	14	4	12	8	22	22	36	29	59	112	36	54	425	100.0		

Source: Survey conducted by authors of book for this study.

TABLE 2-3

PRICE SURVEY OF SERVICE STATIONS LOCATED THROUGHOUT ST. LOUIS—FALL 1969[1]

									Independent Reference				−1¢	Major Reference	+1¢				
Regular	28.9¢	29.9¢				30.9¢		31.9¢		32.9¢			33.9¢	34.9¢	35.9¢	36.9¢	Stations by Brands		Pricing Below Refer-ence	Posting Prices on Signs
Brand Premium	30.9¢	30.9¢	31.9¢	32.9¢	33.9¢	31.9¢	32.9¢	33.9¢	34.9¢	34.9¢	35.9¢	36.9¢	37.9¢	38.9¢	39.9¢	40.9¢	No.	%		
American													6	12	84	5	107	19.6	16.8%	3.7%
Shell												6	7	8	38	2	61	11.2	34.4	19.3
Mobil												2	1	13	39		55	10.0	29.1	1.8
Phillips												2	2	6	40	3	53	9.7	18.9	5.7
Texaco												1		9	32	1	43	7.9	23.3	9.3
Gulf													3	12	26	1	42	7.7	35.7	7.1
Sinclair												2	1	13	20		36	6.6	44.4	11.1
Conoco												1	2	5	6		14	2.6	57.1	50.0
DX (Freeway)											(3)			3	5		11	2.0	54.5	9.1
Monsanto														1	4		5	.9	20.0	
Skelly														2	2		4	.7	50.0	100.0
Subtotal No.											3	14	22	84	296	12	431	78.8	28.5	10.2
Majors %												(3.9)	(5.1)	(19.5)	(68.7)	(2.8)	(100%)			
Fisher Fleet										1	1		3				5	.9		
Zephyr				1			2		1	6	10		1				21	3.8	19.0	71.4
Fina				1		1	1	2	1	5	8		1	1			21	3.8	28.6	52.4
Mars-Marine		1				3	1			4							9	1.6	55.6	66.7
Star				1			1	1		1							4	.7	75.0	25.0
Site		1	1	1	1	1	1	1	3	1	1						12	2.2	83.3	33.3
Derby						2	3			1							6	1.1	83.3	33.3
Bonafide				2	1				1								4	.7	100.0	100.0
Texas Discount		5				6											11	2.0	100.0	36.4
Others	1	1	1	3			3	2	2	6	2			1	1		23	4.2	56.5	69.6
Subtotal No.	1	8	2	9	2	13	12	6	8	25	22		5	2	1		116	21.2		54.3
Price Marketers %	(0.9)	(18.1)				(21.6)		(12.1)		(40.5)			(4.3)	(1.7)	(0.9)		(100%)			
Total	1	21				25		14		64			27	86	297	12	547	100.0		

Source: Survey conducted by authors for this study.

1. Does not include twenty-three Clark Oil Company Stations which at the time of the survey were selling only premium grade gasoline.

TABLE 2-4

Market Leaders—Stations below Reference and Posting Price

	San Francisco			St. Louis		
	Company	Pricing below Reference	Posting Price	Company	Pricing below Reference	Posting Price
Reference marketers:	Union		4.3%	American	16.8%	3.7%
	Chevron	7.0%	5.8			
Aggressive leader:	Shell	30.9	32.7	Shell	34.4	19.3

certain regions or markets and are, therefore, more vulnerable to upset market conditions.

The important majors, in contrast to the market leaders, generally have thinner market representation, poorer market acceptance, and only around an average share of the market. While the marketing behavior of the important majors is significant, it is not nearly as influential as that of the market leaders. In addition, the percentage of their stations with prices below the reference price is normally much higher than for the market leaders and they are more inclined to post their price. For example, in Washington, D.C. the important majors—Sun, Gulf, Texaco, and Mobil—had from 16.4 percent to 54.0 percent of their stations below the reference price as contrasted with the market leaders, which ranged from 13.7 percent to 15.6 percent of the stations below the reference price. In addition, from 15.1 percent to 49.2 percent of the important majors posted prices while only from 2.7 percent to 8.3 percent of the market leaders posted prices.

The minor majors are those major oil companies with relatively few stations, having generally mediocre to poor local market acceptance, and with a small share of the market. Generally, the majors in this category have the largest proportion of their stations below the reference price. The minor majors in Washington, D.C. are Sinclair, Citgo, Atlantic, Phillips, and Chevron and the proportion of their stations with prices below the reference price are shown in Table 2-1. While in Washington, D.C. the minor majors were found to have a relatively small portion of their stations posting price in comparison to the important majors, the opposite was found to be true in San Francisco and St. Louis. In both these markets the minor majors not only had a larger proportion of their stations below the reference price, but also posting price as one might expect.

The categories into which a major oil company is placed and the pricing behavior that it follows vary greatly for certain majors as can be observed from Figure 2-1. Esso is a market leader in Washington,

FIGURE 2-1

CLASSIFICATION OF MAJOR OIL COMPANY STRENGTH IN DIFFERENT MARKETS

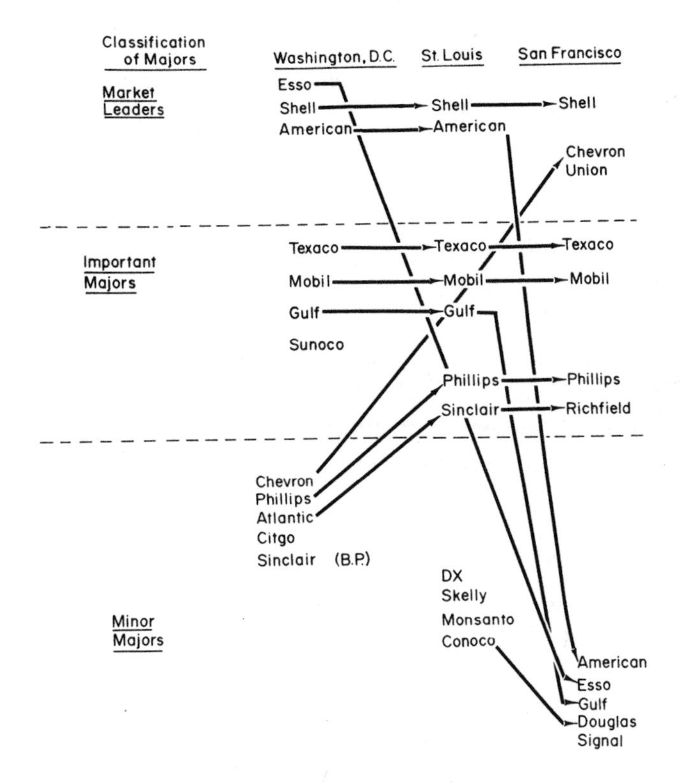

Source: Developed from market surveys conducted for this study.

D.C., has no representation in St. Louis, and prices as a minor major in San Francisco. In contrast, Chevron acts as a market leader in San Francisco, but operates as a minor major in Washington, D.C., and has no representation in St. Louis. American is a market leader in both Washington, D.C. and St. Louis, but prices as a minor major in San Francisco.

Pricing States in Stable and Chaotic Markets

The pricing conditions characteristic of stable as contrasted to chaotic markets can in part be ascertained from studying markets fitting into each of these categories (see Table 2-5). Washington, D.C. and St. Louis at the time studied were relatively stable markets. One of the characteristics of stable markets is that the prices of major-brand gasolines are heavily clustered around the reference price. In Washington, D.C. 56.2 percent of the stations sold gasoline at the reference price and 16.1 percent sold it one cent or more above the reference price resulting in a total of 72.3 percent of the stations selling at or above the reference price. Similarly, in St. Louis 71.5 percent of the major-brand stations sold gasoline at the reference price or above it.

Important majors or minor majors generally had a higher proportion of their stations selling gasoline at least one cent below the reference price than was the case for the market leaders. Market leaders seem to tolerate to a point this minor deviation from reference price because of the poorer development of certain competitive systems and because the one-cent differential frequently makes little difference in the drawing power of the less significant competitors' stations. In Washington, D.C. 20.7 percent of the stations sold gasoline one cent below normal and in St. Louis 19.5 percent sold it one cent below normal. Thus, in both Washington, D.C. and St. Louis more than 90 percent of the major-brand stations priced within one cent of the reference price. This price pattern shows the tight conformance to the reference price that is necessary for stable market conditions, which is a prerequisite for workable nonprice competition.

A second characteristic of stable markets is that the majors generally do not post their prices on street signs so as not to cultivate consumer interest in price. The majors selling at or above the reference price seldom posted prices (see Table 2-5). In fact, if the price postings by the Sun and Gulf stations are not considered only 3 of the 773 stations in Washington, D.C. and St. Louis that were at or above the reference price posted their price on street signs. The reason for temporarily eliminating Sun and Gulf from the analysis is that these brands frequently post their regular and subregular (Gulftane and Sun 190) prices to communicate the lower price of the tane and not the value of the regular-grade gasoline.

TABLE 2-5

Price Surveys of Two Markets with Stable Prices and One with Unstable Prices

Washington, D.C.[1]

	Price Regular Gasoline	Major Brand Stations Number		% Posting Price	Price Marketers Stations Number		% Posting Price
1¢ above:	36.9¢ and over	104	16.1%				
Reference:	35.9	361	56.2	8.9%			
1¢ below:	34.9	133	20.7	31.6	1	1.9%	
	33.9	31	4.8	58.0			
	32.9	11	1.7	63.6	2	3.7	50.0%
	31.9				17	31.5	82.4
	30.9	3	0.5	33.3	26	48.1	84.6
	29.9				8	14.8	87.5
		643	100.0%	15.5%	54	100.0%	81.5%

St. Louis, Missouri[1]

	Price Regular Gasoline	Major Brand Stations Number		% Posting Price	Price Marketers Stations[2] Number		% Posting Price
1¢ above:	36.9¢ and over	12	2.8%				
Reference:	35.9	296	68.7	2.6%	1	0.9%	
1¢ below:	34.9	84	19.5	13.1	2	1.7	
	33.9	22	5.1	59.1	5	4.3	
	32.9	17	3.9	85.7	47	40.5	68.1%
	31.9				14	12.1	50.0
	30.9				25	21.6	28.0
	29.9				22	19.0	77.3
		431	100.0%	10.2%	116	100.0%	54.3%

Detroit, Michigan[3]

	Price Regular Gasoline	Major Brand Stations Number		% Posting Price	Price Marketers Stations[4] Number		% Posting Price
1¢ above:	38.9¢	150	5.3%				
Reference:	37.9	1,283	45.3	4.1%			
1¢ below:	36.9	376	13.3	10.4	3	1.6%	
	35.9	265	9.3	41.1			
	34.9	220	7.7	55.5	5	2.8	40.0%
	33.9	272	9.6	81.2	2	1.1	100.0
	32.9	198	7.0	80.8	46	25.4	84.8
	31.9	47	1.7	78.8	83	45.9	88.0
	30.9 or less	24	0.8	58.3	32	17.6	81.2
	29.9				7	3.9	100.0
	28.9				2	1.1	100.0
	27.9				1	.5	100.0
		2,835	100.0%	26.6%	181	100.0%	83.9%

1. Surveys completed early in the fall 1969.

2. Does not include twenty-three Clark stations since at the time of the survey Clark did not sell regular-grade gasoline.

3. Price survey carried out by the Kent Marketing Service on March 30, 1970.

4. Does not include 117 Clark stations since at the time of the survey Clark did not sell regular-grade gasoline.

The less than 10 percent of the major-brand stations in Washington, D.C. and St. Louis that were more than one cent below the reference price represents a potential threat to market stability and it is important to understand what is happening to appreciate major oil-company price behavior. Generally, the major-brand stations in St. Louis and Washington, D.C. with prices more than one cent below reference were located in pockets of intensive price fighting and it is among such stations that the posting of price by majors is quite common (see discussion of zone pricing in Chapter 7).

Major-brand stations with these unusually low prices are of two types—they are either "maverick major-brand dealers" who consistently try to be more than one cent below the prevailing major-brand price or they are those majors who cut price defensively. The defensive price cutters are often the "aggressive market leaders" such as Shell. In essence, the aggressive price cutters communicate the message, "You come up near the reference price, or I'll come down where it will be difficult for any of us to make profit from marketing gasoline." Furthermore, the research carried out in Washington, D.C., St. Louis, and San Francisco indicated that more frequently than majors policing majors, the aggressive market leaders police the private branders and price marketers. In those areas where the private branders and price marketers were concentrated, or near an isolated private brander, the majors had the largest portion of their stations that were more than one cent off pattern. It appears that aggressive pricing by majors is not a normal practice, but a disciplinary practice to reduce the profitability of price discounting and to discourage competitors—be they major or private branders—from using price to gain a competitive edge.

By way of contrast, Detroit is an example of an unstable market where major branders were having trouble establishing a reference price to which most of them would conform. The reference price at the time of the survey was 37.9¢ per gallon for regular gasoline. Only 50.6 percent of the stations in Detroit were at or above the reference price, while for Washington, D.C. and St. Louis more than 70 percent were at or above the reference price (see Table 2-5). Only 63.9 percent of the stations in Detroit were within one cent of the reference price in contrast to more than 90 percent in Washington, D.C. and St. Louis. Thus, more than one-third of the stations in Detroit were pricing gasoline at least two cents under the reference price and acting as a force to pull the entire market down. In addition 26.6 percent of the major-brand stations in Detroit were posting prices which is close to twice the percentage posting prices in Washington, D.C. and St. Louis.

Not unexpectedly shortly after the survey was taken the Detroit market collapsed. The reference price in less than one month fell eight cents per gallon as efforts were made by the market leader to correct a "bad maverick major problem." Furthermore, the market has and will probably continue to follow a "yo-yo" pattern as the market leaders—American and Shell—try to correct the off-pricing pattern of major-brand competition. Once before as a result of this problem Detroit earned the title of "Worst U.S. Market" and it may very well do so again unless conformity to the normal reference price is accepted by most major-brand competitors.[1]

Brand Image and Product Differentiation

Given the structure of the industry it is understandable why the majors steer away from active price competition and substitute in its place a strategy of nonprice competition. A fundamental characteristic of this nonprice strategy is for major oil companies to see that their gasolines are not classified and sold to the public as commodities where price is often the most important decision variable. The major oil companies do this by branding their gasolines and then by promoting their brand images which result in some customer preference. The major oil companies expend effort and money to brand their gasolines to increase their net realization from sales over what they would be if their gasolines were sold in the unbranded market. Since the return from this strategy has generally proven to be good, the majors have invested considerable money over the years in trying to develop and sustain some preference for their brands of gasoline to the exclusion of gasoline sales on an unbranded basis.[2]

Developing a strong brand preference is a very difficult task since gasoline is a liquid product which hopefully customers never see, smell, or taste. The branding of gasoline parallels the problem of trying to brand electricity, natural gas, or water. As a result while a good brand image is important to a major oil company it is not by itself a sufficient reason for a driver to purchase a given brand. However, a reasonably well-recognized brand is necessary before the other elements of the nonprice competitive strategy can be used effectively.

Many elements enter into the image that the driver forms of a particular brand. Some are psychological in origin and difficult to assess. But this does not make them any the less real or important. Others are clearly aesthetic in character such as station design and appearance, color and emblems. Still others derive from the character and efficiency of the retail operation. Poor retail operations suggest sloppy quality control at manufacturing. One of the important parts of the marketing strategy followed by the majors is symbols. Logos are frequently changed in efforts to update a brand's image in order to communicate a sense of progressiveness and modernness.

Brand image and the consequent brand preference are importantly shaped by advertising. When the product itself is impossible for most customers to evaluate as it is with gasoline, slogans and symbols play an important role in the building of a producer's image. As an example, one of the more successful advertising campaigns in the gasoline business is Humble's "Put a Tiger in your Tank." The connotation is, of course, power, but it is doubtful that this brand of gasoline is significantly more powerful than most other gasolines. American Oil's "You expect more from Standard and you get it" has also been an extremely successful campaign. Like the Humble "Tiger" it says nothing concrete about the product, but is concerned exclusively with image building.

Some of the major oil companies take a more direct approach to image building by trying to overcome the notion that most gasolines are basically alike. One strategy is to build an image around the technical aspects of gasoline, which are impossible for most drivers to evaluate. The classic example of this strategy is Shell's platformate campaign. Actually, the platformate advertising is on a par with a bakery's claim that its bread contains flour. Most gasolines contain platformate or its equivalent. A similar approach is for majors to try to differentiate their gasolines by chemical additives such as TCP (now TCP 2), and Boron. While these additives do improve the quality of gasoline, most competing brands have their own additives, which result in there being little nor any real difference in their quality. Apparently, such strategies work. For over the past ten years Shell has been plugging its

TCP, platformate, and "special mileage ingredient" and has seen its position rise from sixth place in 1960 to second place in 1969.[3]

To counteract the superiority claims of certain major oil companies about their gasoline the American Petrofina Company—Fina brand—has followed a rather unique approach. The Fina strategy has been basically to "poke fun" at the competitive claims. One of their campaigns in the early 1960's promoted Fina's secret ingredient. Fina claimed that their gasoline was loaded with all the traditional additives which made it exactly as good as the best. However, their ads solemnly declared that Fina-brand outlets were the only stations with the ultimate ingredient—pink air. "It was colorless, odorless and invisible, but it was great." American Petrofina followed with their "Pflash Campaign" and portrayed pflash as their antinuisance additive.

> Pflash combats all those petty driving pfrustrations that pfatigue you, inpfuriate you, and pfinally make you sorry you ever got a driver's license.
> Pfabulous as it may seem Pflash makes even parallel parking a cinch. It smooths out bumps . . . firms up soft shoulders, adds minutes on parking meters . . . and even improves the food at roadside restaurants. . . .[4]

Some of the testimony given during the 1965 Federal Trade Commission's Hearing on the Marketing of Automotive Gasoline pertained to the similarity and differences among major-brand gasolines. The following is part of the testimony given by Stanley Learned, president of Phillips Petroleum Company, in answers to questions by Rand Dixon, chairman of the Federal Trade Commission.

Chairman Dixon:	Do you exchange any gasoline (trade with other majors) in the East?
Mr. Learned:	We have some we exchange, yes.
Chairman Dixon:	That makes it hard to call yours Phillips, doesn't it?
Mr. Learned:	As long as they meet our specifications.
Chairman Dixon:	You put a little pinch of something in it that makes it a little different?
Mr. Learned:	Yes; we have an additive which allows us to advertise. *I don't know whether it does anything for gasoline.*[5] [Emphasis added.]

The antiknock property of gasoline, which is measured by its octane rating, has resulted in still another product strategy followed by the industry. This is the division of gasoline into five basic grades—subregular, regular, middle, premium, and super premium (see Table 2-6).

TABLE 2-6

NUMBER OF GRADES OF GASOLINE GENERALLY SOLD BY MAJOR OIL COMPANIES IN 1970[1]

Rank on Assets	Company	Sub-regular	Regular	Middle Grade	Premium	Super Premium	Number of Grades
1	Standard Oil N.J.		X	X	X		3
2	Texaco		X		X		2
3	Gulf	X	X		X		3
4	Mobil		X		X		2
5	Socal						
	Western Division		X		X	X	3
	Eastern Division	X	X		X		3
6	Standard Oil Ind.		X		X		2
7	Shell		X		X		2
8	Atlantic Richfield		X		X		2
9	Phillips		X		X		2
10	Continental	X	X	X	X		4
11	Sun Oil	X	X	X	X	X	8
12	Union-Pure		X		X		2
13	City Service		X		X		2
14	Getty				X		1
15	Standard Oil Ohio		X		X		2

1. Low-lead gasoline began to appear in the later part of the year.

The recent development of the blending pump now allows eight different levels to be sold economically. With the exception of the recent move by Getty, which sells only premium, all of the majors offer at least two grades—regular and premium. Five of the fifteen largest oil companies market three or more grades. Gulf, Socal (Eastern Division), Continental, and Sun sell subregular grades of gasoline. Esso, Continental, and Sun sell middle grades, and Socal (Western Division) and Sun both sell super premium grades of gasoline.

The octane requirement of cars in operation in 1970 for which the various grades of gasoline have been developed is shown in Table 2-7. In the fall of 1969 the Federal Trade Commission held hearings on the posting of gasoline octane on the pump. Without exception, the majors opposed this proposed trade regulation. The limited scope of their testimony seemed to reinforce the idea that the major-brand oil companies prefer to market their products on image differences rather

TABLE 2-7

RESEARCH OCTANE NUMBERS RELATED TO AUTOMOBILE REQUIREMENTS

Research Octane Numbers	Percent of Automobiles Satisfied
88	10
90	20
92	40
94	60
96	80
98	90
100	98

Source: Oil Daily, *March 3, 1970, p. 6.*

than on technical specifications. Yet, the fact remains that the major oil companies exchange gasoline among themselves at agreed specifications but are unwilling to sell gasoline to the final customer in this objective manner.[6] Furthermore, the gasoline sold at one major-brand station is frequently the same as that sold under a different brand. In a recent state survey, Sinclair was found to be selling gasoline produced by Pure, Marathon, and Signal in addition to that which Sinclair produced itself. A Humble station was selling gasoline supplied by Triangle Refineries that also supplied certain Citgo stations.[7] In the case of "common stream" pipelines such as Williams Brothers, gasoline meeting certain specifications is received at one end of the pipeline while the shipper receives the right to draw out this amount at the other end. The identity of the specific shipment is not maintained.[8]

A new dimension in the product strategy of the major oil companies surfaced during 1970. In January Edward N. Cole, president of General Motors, declared that the major oil companies must "get the lead out" of gasoline; this position was rapidly endorsed by Henry Ford II. By the end of 1970 most of the major oil companies had introduced low-lead/no-lead gasoline which was recommended for the 1971-model cars and which supposedly will be required for all new cars in the future. The advent of the low-lead/no-lead gasoline has resulted in additional oil companies such as Texaco and Shell going to three grades of gasoline. The new gasoline has become an important tool for developing the distinctiveness of major-brand gasolines. For example, American Oil is promoting its no-lead Amoco gasoline with its "Fresh as a Daisy" theme, while Shell is telling drivers to buy "Shell of the Future" at its branded stations.

Despite the fact that major-brand gasolines, physically at least, are not very different, or are not different at all, efforts at product differentiation remain the cornerstone of the major firm's marketing strategy. All other elements of their marketing strategy depend upon effective branding which results in widespread acceptance of brand names as symbols of quality products sold by progressive and reliable companies.

CONTROLLED SINGLE-BRAND SERVICE STATIONS

THE DRIVE-IN SERVICE STATIONS selling a single brand of gasoline, complementary products, and a variety of car services now seem to be the obvious way for marketing major-brand gasoline. However, gasoline and related products were not always sold in this manner. Before 1910 gasoline was sold by livery stables and garages and by hardware, grocery, and general stores. As a result of the dramatic growth of automotive registrations from 312,000 in 1909, to 7,565,000 in 1919, to 26,501,000 in 1929, there emerged a clear need for a new type of outlet for marketing gasoline.[9] The radical departure was the service station, which specialized in handling the soaring demands for automobile gasoline and lubricants and car services.

While the need for a new kind of retail outlet is now obvious, the reason the major oil companies decided to own or otherwise control these outlets is less clear. The initial demand for service stations was quickly met by thousands of independent businessmen who were eager to build outlets to take advantage of the good profit opportunities that presented themselves. During the mid-twenties and early thirties there was intensive competition among the majors for exclusive representation by the new, high-quality, independently owned retail outlets which made them particularly profitable. Since the major oil companies found themselves in a position of costly bidding for the quality outlets they gradually integrated forward to own their key marketing facilities or control them through long-term leases.[10]

A problem related to the intensive competition for representation by the quality independent stations was the switching of brands, which disrupted major oil company representation in an area or a region. Switching was relatively easy for the independent operators. In most cases all switching required was for the owner of a service station to allow his inventories to run low, to change brand signs and symbols, and to order his product from a new supplier.

A further reason the majors sought to own their outlets was so that they could control in detail the character of the retail operation and retail setting within which their gasoline was sold. Such control is important in the development of a favorable brand image. In other businesses such as the retailing of automobiles or quick food sales either good alternative sources of supply are quite limited or the cost of changeover too high to permit dealers to change brands. Thus the suppliers can exercise control over the internal operations of their dealers without owning the operation. The threat of cancellation or nonrenewal of the supply relationship is enough. However, in the case of gasoline, ownership control was needed to bolster the threat of cancellation or nonrenewal of the supply relationship since there were many equally good suppliers who would jump at the chance to snap up a competitive dealer who controlled a quality location.

Another aspect of the strategy of the major oil companies has been to sell their gasoline and oil products through outlets specializing in sales of a single brand—exclusive product outlets. This practice reinforces that of product differentiation and the branding strategy of the major oil companies. Not only are products advertised as being different, they are also sold through different outlets. In addition, the exclusive product outlets help the major oil company to control the activity of the dealer who has only one supplier and as a result almost no bargaining power. Another reason for exclusive product outlets is to reduce the opportunity that customers have to make direct price comparisons. However, gasoline has not always been sold on an exclusive basis and occasionally still today a driver will accidentally come upon a service station selling multiple brands. In the later 1910's and early 1920's it was not uncommon to find independent service stations selling two brands of gasoline. Even today we might have large-scale, efficient service stations selling multiple brands of gasoline were this not contrary to the interest of the dominant major-brand oil companies. This type of arrangement would be similar to specialty appliance stores where very frequently two or more competitive brands are sold side-by-side and to several other types of mass retailing where customers are given a choice among competing brands.

While practically all major-brand gasolines are sold through exclusive retail outlets, there still remains an important legal question about this practice. Section 3 of the Clayton Antitrust Act makes it unlawful for a company to sell its goods to another for resale if it makes the sale on the condition that the buyer will not handle the products of a competitor where the effect may be substantially to lessen competition. Landmark antitrust suits against Standard Oil of California and Richfield established that "any large supplier of a commodity like gasoline, will not be legally safe if he enters into exclusive agreements of any kind with his distributors." [11] On the other hand "the Standard and Richfield precedents do not, and in actual practice, cannot prohibit a voluntary choice of exclusive handling by dealers who are offered substantial freedom of choice to do business in other ways." [12]

As a consequence, the majors have been very careful to word their leases and other formal contracts with dealers to state clearly that no operating control, including exclusive dealing, is an expressed or implied condition of the agreement. The fact remains that either as a result of subtle threat, coercion, the structure of the system, or by voluntary acceptance most stations exclusively sell only one brand. Attempts by independent service station owners to sell more than one brand to strengthen their competitive position occurred during the early days of fair trading in New Jersey.* In response to this development J. N. Carney, general sales manager of American Oil, said that they had two or three cases of split-pump operations shortly after the resumption of fair trading: "We gave them ten days' notice, then moved in and pulled our pumps. It seemed to have a good morale effect on the rest of our dealers." [13] The report also stated that Shell had similar cases and broke its contracts with those involved. In such cases the pressures for exclusive dealing were quite overt while normally they are much more subtle.

INTENSIVE MARKET COVERAGE

THE COMPETITIVE APPROACH of the major oil companies so far developed is a nonprice strategy, built around branded gasoline sold through outlets handling a single brand. Given this marketing orientation, what type of service station coverage of markets would seem to be appropriate? Would it be more advantageous for branded

* In states having fair trade laws manufacturers are permitted to establish minimum prices at which their products can be sold.

gasoline to be sold (1) through relatively few, large-volume, centrally located stations, or (2) through large numbers of very accessible, highly convenient, and relatively low-volume stations? The general strategy has been one of intensive development of the market through large numbers of highly convenient stations. For example, some extremely successful major brands have close to a thousand exclusive outlets in a single metropolitan market. Market coverage is so crucial that well-known major brands that are lightly represented in particular markets fare quite poorly, though they are dominant and quite successful elsewhere. American Oil sells through nearly 30,000 branded outlets as does Humble, and each is dominant in certain major markets. However, both are weak in California and Humble has never made it in Chicago. On the other hand, the number one gasoline marketer, Texaco, sells through more than 40,000 branded outlets while Shell, the up-and-coming number two marketer, sells through only 22,000. Yet Texaco is only slightly larger than Shell in gasoline volume (8.7 percent vs. 8.5 percent share of the market in 1968 according to *Fortune* magazine). The connection between sales volume and market coverage is not perfect, but it is very strong. The confounding elements are quality of location and physical facilities.

Earlier in the history of the gasoline industry the connection between market coverage and sales volume was much clearer, and the strategy of all majors was to get as many stations as possible of reasonably good quality. Gasoline was indeed a convenience good, sold through convenience type outlets along traditional friendship lines. There is a close parallel between the neighborhood service station and the "Ma-and-Pa" grocery operation. But this is now changing. Just as important economic and social forces have all but eliminated the "Ma-and-Pa" grocery, so have forces been operating to change the character of gasoline retailing, but not quite so dramatically. There are several reasons for this, but perhaps the most important is the continuing dominance of convenience as a patronage motive in purchasing major-brand gasoline. This is especially important when the majors' aversion to price competition prevents the savings of the supermarket-type operation from being shared with the customer in the form of a price cut. It remains to be seen how much the motorist really values convenience were it possible to buy major-brand gasoline for less from super stations. The experiences of price marketers, which are described in the next chapter, suggest that convenience is important primarily because the major-brand customer has to pay for it whether he wants it or not—so why not?

Still, even without the accelerating effect of lower price, the retail structure of the major-brand operation is moving inexorably in the direction of increasing average volume through fewer and large stations. The result is a collage of convenience and high-volume stations with the predominant emphasis still on convenience, but with growth primarily in the high-volume type station. This means that location looms increasingly important in the retail strategy of the majors and proportionately more money is being spent on location and facilities than in the past.

The forces at work that are bringing this to pass are reasonably clear. They are less social pressures from the demand side than they are economic pressures from the supply side. It is becoming increasingly difficult for a dealer to make a living from one of the lower-volume stations. Dealer turnover of stations owned or controlled by majors was 24.6 percent in 1970, but was much higher for the lower-volume stations as shown in Table 2-8. The result is that it is impossible to keep many of the lower-volume stations open.

TABLE 2-8

Service Station Turnover Related to Volume of Business During 1970

Annual Station Volume	Number of Stations	Number of Terminations	Turnover
Less than 100,000 gallons:	9,615	3,316	34.5
100,000–200,000 gallons:	21,539	6,971	32.4
More than 200,000 gallons:	53,568	10,870	20.3
Total:	84,722	21,157	24.7

Source: National Petroleum News Factbook, *Mid-May 1971, p. 112.*

The super station was pioneered by the price marketers and then adopted by the major oil companies (e.g., Standard of California bought the Urich operation and Humble purchased Oklahoma). However, the economic facts of life made it necessary for the majors to adapt the super station idea to fit their needs. The trick was to get needed volume without discounting. The answer has been to place stations in premium traffic locations and to invest heavily in elaborate facilities.

For major-brand super stations to generate volume sales without discounting they have to have certain attractive features to compensate for their being less convenient than the neighborhood stations. The attractiveness of the super stations is that they are big and impressive, which means, among other things, that they are often modern, efficient looking, spacious, easy to enter and leave, well lighted, and open long hours. However, such stations require a lot more money. American Oil estimates its average new station costs $170,000, with $65,000 of this being the cost of the land by itself.[14] Contrast this with the 1952 figure of $50,000, only $20,000 of which was for land.[15]

Beside the fact that convenience is still a patronage motive and that the major super stations are costly to operate and do not generally cut price, there is still another reason why the super station movement is so gradual. The answer seems to lie in the fact that obsolete stations continue to hang over the market much longer than the small retail grocery store did. The major oil company which owns the lower-volume station frequently encourages its dealers through a variety of practices to continue operation of such stations even though it knows that in the aggregate similar policies followed by all majors will increase the cost of distribution and stunt the growth of their new super stations. Yet, each major firm feels some pressure to act this way for it knows that if it closes a station selling less than 20,000 gallons a month, only a fraction of this business will shift to its other stations and that much of the business will go to competitors. To keep the station open reduces the volume that its more efficient stations could be doing. But the loss is shared by competitors as well as by its own stations. On the other hand, should a company close such a station, it bears the full brunt of the loss, while receiving only a fraction of the gain. Hence, major oil companies are often motivated to keep declining stations open as long as they can. This situation has its parallel in the old parable called the "Tragedy of a Commons." In this parable, each herdsman knows that adding one more sheep to the flock grazing on a bounded common meadow has two components.

> The positive component is a function of the increment of one animal whose proceeds, when sold, will accrue exclusively to the individual herdsman. The negative component is a function of the additional overgrazing created by one more sheep in a bounded meadow. But since the effects of overgrazing are shared by all herdsmen, the negative utility for any particular herdsman is only a fraction of its total disutility. The result is that each individual herdsman concludes that the only sensible course for him to pursue is to add an-

other sheep to his herd; and another. The same decision is reached by each and every herdsman sharing a commons. The result is ruin to all.[16]

In a sense the pattern of gasoline retailing reflects this parable. Over the years the major oil companies have followed the practice of blanketing markets with outlets to gain and retain a desired share of the market. As a consequence, the costs of distribution in this industry are very high and because of low utilization of retail capacity, a very large number of dealers are barely making ends meet. This situation is one of the more unhappy consequences of nonprice competition in this business. It represents one of the grosser misallocations of land, labor, and capital in our society.

This problem has not gone unrecognized by the major oil company executives. For example, D. Woodson Ramsey, vice president of marketing, Humble Oil (the domestic marketing arm of the world's largest oil company) stated: "The industry has more damn gasoline stations than we have customers—the oil industry problem is to find some way to deliver gasoline to the consumer other than by having service stations on every corner. It is too expensive." [17]

In 1970 the chairman of the board of Standard Oil of New Jersey, Ken Jamieson, said when talking about the service station problem: "We are too standardized on every street corner, with the same structure and same merchandise. That isn't the answer." [18]

The statement of Tom W. Sigler, vice president of marketing, Continental Oil, in 1970 reinforces what has been said about station overbuilding:

> It has been frequently charged in the past years that the belief prevailed among some corporate managements that typical two bay stations on every corner were necessary in order to market crude —without regard for the profitability of the marketing investment. . . .
>
> The attempt to push crude through stations has been at least partially responsible for one of our greatest problems—that of too many stations. In my opinion, the conventional service station representation in many parts of the country is much greater than is needed to serve the consumer public. It's a case of marketing "overkill." And it is an indication that we have been subsidizing unprofitable outlets rather than facing the music and closing them down. Then we build a raft of new stations in markets that may already be

heavily overbuilt, the available business is spread so thin that few outlets really have enough volume to justify their existence.
We can, however, do a much better job of identifying those stations that should already be closed down and not hang on to them with nothing but a hope, a prayer, and a subsidy. And we can do a better job in recognizing an overbuilt situation and not blindly add to the problem with more new outlets like straws to a camel's back.[19]

Patronage Tools

A problem associated with the nonprice competitive strategy of the major oil companies has been that most gasoline customers are not committed to purchase a particular brand of gasoline because of the close similarity of most major-brand operations and the large number of different major-brand stations available. Such conditions make it easy for gasoline purchasers to switch their purchases from one major station to another, or to divide their purchases among several different brands. To reduce this problem the major oil companies sought ways to tie otherwise indifferent customers to particular brands of gasoline. To force more customer commitment they have employed a number of costly patronage-building tools, including credit cards, trading stamps, and continuity premiums. Through these devices the major oil companies hoped to insure for themselves a large portion of the gasoline and other automotive purchases of given customers.

Credit Cards

In certain respects the use of credit to sell gasoline seems rather strange. Individual gasoline purchases involve relatively small amounts of money spent to purchase a product that is quickly used up. Furthermore selling gasoline on credit does not expand the total demand. All these factors argue against the use of credit to sell gasoline. These are much the same reasons credit is not used to sell groceries. Credit is normally employed to assist shoppers in buying big ticket durable goods—items with a relatively long life span—and purchases that would not in many cases otherwise be made or else would be substantially delayed. Thus, it is clear that credit is primarily used in the oil industry as a marketing tool to increase brand loyalty.

In the latter 1940's credit-card sales of gasoline were very restricted and the cost of such sales was quite high. Various estimates placed the direct cost of credit sales as being upwards to two cents per gallon of gasoline, which was large relative to other marketing costs at that time. As a result, only around 5 percent of gasoline was sold on credit. At that time, the credit card was a prestigious "courtesy card," issued to only a small and elite portion of the market.

In the early 1950's as supply and demand conditions came more into balance, the oil companies increasingly turned to credit as one of their competitive tools. Throughout much of the 1950's credit-card sales increased at a rate of 20 to 25 percent a year.[20] The growth of credit cards continued through the 1960's. Industry officials estimate that credit sales increased from $4 billion in 1961 to $7 billion in 1966. In 1970, annual credit sales were estimated to be near the $10 billion level through approximately 100 million credit cards. The spurt in credit-card sales in the latter part of the 1960's resulted from mass mailings of unsolicited cards. The relative importance of credit sales is indicated by the 1966 estimate that 70 percent of West Coast sales were by credit card, while figures for the East Coast were running at about half that rate.[21] Shell Oil stated in the latter part of 1969 that 37 percent of the gasoline it sold was on credit.[22]

The actual and complete cost of credit sales of gasoline appears to be a closely guarded competitive secret. However, the real cost of gasoline credit-card sales has to be relatively high because of the nature of the transaction—large numbers of small transactions resulting in high handling and processing costs. For example, oil-company credit sales average $6 to $7 in comparison to average bank card sales of from $17 to $25. The credit departments themselves become sizable aspects of major oil companies' operations—as one case, a major oil company in 1964 was reported to have a major IBM computer system and 650 people to run its credit operation.[23] A public statement by the Mobil Oil Company in 1968 put its credit-card system cost in excess of $20 million a year.[24]

Directly affecting the cost of credit transactions are the costs of issuance of credit cards, handling and processing costs, bad debt loss, and the cost of capital that is tied up for thirty to sixty days. The cost of issuing credit cards runs from one dollar for an unsolicited card to five dollars for a solicited card with a credit check. Studies of processing and handling costs for gasoline credit tickets indicate that they run about 2.5 percent of sales. Indications are that bad debt losses might easily run 0.5 percent. The cost of capital tied up for thirty days without charge might cost another 0.8 percent. As a result of such costs it is not hard to arrive at an estimate of 4 percent cost of gasoline credit-card sales and possibly up to 5 percent or more as certain companies estimate to be a realistic figure. These costs do not include the indirect-handling cost of making a credit sale at the station level.[25]

Using the 4 percent figure and the approximate cost of a gallon of major-brand gasoline at 37.9¢ (35.9¢ for regular and 39.9¢ for ethyl), the cost of making a credit-card sale runs approximately 1.5¢ per gallon of gasoline (37.9¢ × .4 = 1.52¢).

The cost of gasoline credit sales could be significantly decreased if the oil industry were to adopt a universal or widely accepted credit card. In 1952 a national credit-card plan was developed, but it failed without gaining the support of a single major oil company. During the middle 1950's a major oil company executive suggested that approximately 50 percent of the cost of credit could be saved if the industry adopted a universal credit card. Nothing happened as a result of this idea. In the latter 1960's the bank card systems—now Bank Americard and Master Charge—became available to the major oil companies, but they have not been widely accepted and many of the market leaders honor only their own company credit card.[26]

The reasons the oil companies have continually resisted the general credit-card schemes is that they would lose one of their strongest marketing tools which bind customers to particular brands of gasoline. While cost considerations do enter the picture, under conditions of nonprice competition, these additional costs are not as important to many companies as the loss of control over their existing credit-card customers. However, with the growing acceptance of the bank cards, the oil companies have had to strengthen their credit cards. Today with most oil credit cards, a driver can charge his food and lodging at certain motel and hotel chains. In addition, the oil companies are making available premium merchandise to their credit-card holders on easy time-payment plans to encourage them to use their cards.

Credit cards, similar to station building, has become an overused competitive tool. While the industry is plagued with too many multiple credit-card holders,[27] the individual oil company still finds it beneficial to issue more credit cards to attract customers to their individual brand systems. The result is a growing number of costly, inactive credit cards; some companies are reporting that 50 percent of their credit cards are not used even once a year. The Federal Trade Commission rule banning the unsolicited mailing of credit cards effective May 17, 1970, helped the oil industry to correct a problem that seemed to be growing out of hand.

Trading Stamps

SOME PEOPLE have several oil-company credit cards and others don't use credit cards. Thus, an opportunity exists to try to restrict the switching of those gasoline purchases that are not strongly influenced by a single-brand credit card. The long-run promotion that complements credit cards is to lock customers into a system of trading stamps.

The trading-stamp principle is a little different from the credit-card approach. The credit-card system is owned and operated by the oil companies and is designed primarily to benefit them. However, the trading-stamp system is not owned by the oil companies and, furthermore, the supermarkets and not the service station are the big-volume stamp givers and the anchor of the system.

The principle of the trading-stamp concept is to lock shoppers into a system of stores giving a particular stamp. A service station adopting a stamp that is important in its trade area then benefits by the semiexclusivity of noncompeting stores giving a particular stamp. Those gasoline purchasers who are disposed to collect a particular stamp are going to be tied to stations offering that stamp. Thus stamps become another marketing tool for increasing commitment of certain drivers to a particular brand of gasoline.

Trading stamps also give their users an opportunity of occasionally promoting with stamps. Often service stations can promote with double stamps one or two days a week without upsetting a market. When a dealer gives double stamps he is in essence temporarily reducing his prices by approximately one cent per gallon of gasoline. In contrast, if a dealer were to reduce his price by one cent per gallon and put out a street sign, he would likely cause a price reaction.

During the trading-stamp boom of the second half of the 1950's and the early 1960's service stations moved rapidly into the business of distributing stamps. The magnitude of the practice of giving trading stamps is indicated in the market surveys carried out. In St. Louis 66 percent of the major-brand stations surveyed were giving stamps; in San Francisco, 64 percent; and in Washington, D.C. (a weak stamp city), 41 percent.

As with the other nonprice competitive tools, trading stamps are rather costly. The national stamp plans, such as S&H and TV stamps frequently cost service-station dealers between 2.5 and 3.0 percent. This results in a cost of about 1.0¢ per gallon of gasoline (37.9¢ × .025 = .95¢), when stamps are given at the normal rate of distribution (ten stamps for every dollar of purchase).

Short-run Continuity Promotions

Credit cards and trading stamps are long-run continuity promotions that are in large part designed to lock gasoline customers into a brand system. Beside the long-run type promotion, short-run continuity promotions are needed to attract new customers into the system to make up for the natural attrition of customers. In addition, this type of promotion makes purchasing of gasoline more interesting for existing customers. If a new customer becomes interested in a promotion he will likely trade with the station as long as it is necessary to complete it. Commonly, promotions are designed to keep customers coming back for eight to thirteen weeks. Hopefully, by the end of that time the new customers will have been sold on the other attributes of the station and remain its regular customers. Some of the more common types of short-run continuity promotions include drinking glasses, mugs, steak knives, flatware, Melmac, and china.

One of the more successful short-run continuity promotions ever used has been games of chance. The industry practice of using games started with Tidewater on the West Coast. During the first half of 1966 Tidewater used a game called Win-A-Check and recorded monthly sales gains of up to 50 percent.[28] Tidewater's experience sparked widespread adoption of games.

During 1967 and 1968 games were used extensively in the gasoline business. Those companies that were employing the better games were able to temporarily increase their market shares. For example, the 1,931 Shell dealers in New York and Los Angeles participating in the Mr. President coin game increased their average sales from 39,492 gallons per month to 51,567 gallons per month during the period of the game.[29] The temporary sales gains from games is illustrated by

Sun Oil's experience during the three months of its game "Sunny Dollars" (see Figure 2-2). The extent to which games and other short-

FIGURE 2-2

Sun Oil Company's Market Gain During Three Game Months

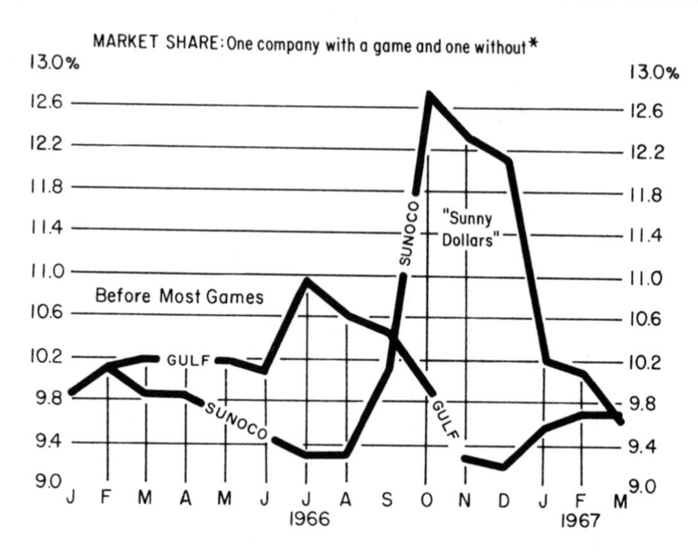

Source: National Petroleum News (*July 1967*), *p. 68.*

* Share of total gallonage for 15 major marketers in New Jersey.

Over a 15-month period from January 1966, to March 1967, Gulf and Sun started and ended with just about the same market share in New Jersey. But during the period, they moved dramatically in different directions. Sun, benefiting from a three-month game, moved up sharply during the game period. Gulf, with no promotion, hit a market-share low while others were playing games.

run promotions are frequently used is illustrated by the experience in Boston and Chicago during the second half of 1968. Within a six-month period, nine competitors in Boston and eight in Chicago were using short-run continuity promotions. With so many competitors simultaneously using games and other promotions the general effectiveness of these tools diminished.

Several of the companies using games would have preferred rather not to have employed this controversial promotion tool, but found that they were forced to do so. During the House of Representatives' Committee on Small Business' and the Federal Trade Commission's hearings on games, several major oil company executives indicated that they were opposed to using games, but found that they had no choice in the matter. Once again, the industry was helped out of its problem when the FTC proclaimed trade regulations on the use of games that became effective in the latter part of 1969. The difficulties in conforming with the regulations have discouraged many companies from using games. In addition, the nature of the new rules which an oil company must follow while employing a game has reduced the customer appeal and, therefore, the effectiveness of games.

Since customers are relatively uninterested in gasoline as a product, much of the brand-name advertising is associated with the periodic promotions sponsored by the oil companies. For example, during the heydays of game promotions a great deal of advertising effort was placed behind the games and little was said about the products and services of the oil companies. Now that the popularity of games has faded advertising is focusing around other premiums. For example, Mobil's advertising tells drivers to come to their stations to get free thermo glasses, or American is talking about their flatware or china offer.

While the short-run continuity promotions are generally less costly than credit cards and trading stamps because of their intermittent use, they still carry a sizable price tag. During 1967 the fourteen major oil companies spent $77.6 million on games of which they recovered around one-third from the dealer.[30] With the dealer cost being approximately 0.25¢ per gallon, the overall cost of a game probably was about 0.75¢ per gallon during the period of the promotion. Other types of promotions may be more expensive. Mobil estimates dealer costs for glassware to be 0.8¢ per gallon of gasoline.

Physical Facilities

The move to the super station and the need to draw volume without discounting has caused both location and facilities to be emphasized in major company strategies in recent years. Initially all service stations were quite similar. The functional porcelain "box" with one or two service islands and a couple of service bays came to be quite standard in the period between 1930 and 1960 (see Plate 2-1). These stations were distinguished by their signs and color schemes; otherwise they all looked pretty much the same.

In the last ten to fifteen years, conditions have been gradually changing. The first move was to the larger multipump layout with considerable thought being given to pump island position, number of islands, and location and lot size. The purpose was to communicate a new sense of quality and efficiency about the operation. The pioneering independents had already shown that the large, open, multipump facility could draw business. They were in tune with the free-standing, open-shelf, mass-merchandising spirit of the period. Streamlined, efficient looking, impersonal marketing had appeal for major-brand stations even without price reductions. Spacious major-brand stations with three or four pump islands began to appear. While stations were growing much larger during the 1950's the metal and glass box was still the architectural rule (see Plate 2-2).

During the 1960's the major oil companies increasingly discovered the marketing potential of station and location aesthetics. This development, which moved into full swing in the latter 1960's, has been approached in two ways—new locations and stations, or the remodeling of existing facilities. Apparently, the shift in this direction was in its earlier stages a forced type of movement. Shell was the leader in the trend to upgrade station appearance with its ranch style design. In 1957, the city fathers of Millbrae, California, forced Shell to build a station that blended into the carefully planned and landscaped residential area. Three more of the ranch style stations were built over the next five years. However, between 1962 and 1967, 90 prcent of the 3,500 stations Shell built were the ranch style design (see Plate 2-3).[31] In many markets Shell tore down or sold its porcelain boxlike stations, often replacing two lower-volume stations with one of its larger super stations which would do more business than the combined sales of the two smaller ones. While Shell had several years' lead in the upgrading of stations their competitors were scrambling to catch up. Since Shell puts its emphasis behind the "ranch style" station its major competitors generally moved in other, and in some cases, all directions. Texaco is building its "Matawan" stations, Mobil its "Pegasus" stations, and American seems to be favoring its "colonial station."

Mobil built its first "Pegasus" stations during 1966. The design is focused on the tubular pumps covered by a circular canopy thirty feet in diameter. The company feels that the design makes a station that is "alive, but not offensive." Through 1968, 55 of these stations were constructed. By 1971 Mobil projected that 3,400 of its stations would receive the new treatment.[32]

Since the cost of starting from scratch and building the new stations is so expensive many existing stations with good potential are being remodeled to improve their appearance. Remodeling a station can cost anywhere from $15,000 to $50,000 while a new super station is likely to cost from $150,000 to $250,000. Remodeling stations amounts to engulfing the original block station in a new shell. The roof treatment is one of the key aspects in remodeling. The Mansard roof has become a favorite style while others have gone to an angular domestic style roof that may be capped off with a cupola. The sides of the building are dressed down in brick or if the surface material is in good condition it is painted according to the new color scheme.

A tremendous amount of money is being poured into creating a more favorable effect for facilities to sell gasoline. As one competitor capitalizes on the marketing potential of the new station look, others are compelled to follow suit to hold their market positions. While Shell has had the competitive edge in this development others are scrambling to catch up, and a massive drive that is resulting in the "elaboration" of facilities seems to be in the making. The cost of many of the new stations is probably close to four cents per gallon. This estimate is based upon a station which costs $200,000 selling 50,000 gallons of gasoline per month with a 1 percent return per month on the cost of the investment. (The 1 percent per month is a common return expected from investment in specialized retail facilities.) Furthermore, as others catch up in the race for fancier facilities the big gallonage figures that the better appearing stations initially drew are likely to decline. In part Shell's earning problem in 1970 seemed to be associated with the difficulties it had in maintaining necessary station volume so that it could economically amortize the cost of its new elaborate station facilities.

Character of the Retail Offer

The retail assortment and level of service that are offered at the major-brand station is one of the principal characteristics that distinguishes the major from the price marketers and private branders. The product assortment and the services of the major-brand station follow as a consequence of the majors' emphasis on the nonprice means of competition as well as the low volume of gasoline sales which characterize the typical major-brand stations. In most cases the major-brand dealer cannot survive on the profit made from selling gasoline. For the major-brand service station dealers to succeed, a significant portion of their sales has traditionally come from minor automotive repair services and sales of tires, batteries, and accessories (TBA). One of the industry's respected management consultants has suggested that a balanced sales effort would result in sales divided into the following categories:

Gasoline	71%
Oil and lubricant	4
Tires, batteries, and accessories	16
Minor car services	9
	100%

Indeed a study done of one of Esso's sales divisions has shown a strong relationship between a high ratio of nongasoline sales and the probability of a dealer succeeding.

To an extent, however, there is a trade off between nongasoline and gasoline sales. For example, when a dealer overemphasizes repair work the quality of service given at the pump is likely to diminish. It is not easy for service-station attendants to switch from working under the hood to giving quick and courteous service at the islands. Repair service also tends to detract from the general appearance of the station and to crowd the limited space that is available. The layout of service stations so that customers often stare into service bays has not helped the problem.

Some major companies such as American Oil tend to give considerable emphasis to station sales of nongasoline products. They want dealers who will provide a range of high-quality, reliable automotive service to their customers. The extent to which this is emphasized in their program is shown by their development of the American Oil Motor Club which offers towing service, emergency repairs and assistance, and even travel insurance. This service-oriented program is supported by extensive training programs and customer promotion.

TBA sales account for between 50 and 60 percent of nongasoline sales. While this is the most important category of supplemental sales, the major-brand service station has been losing its share of this market. Ten years ago 40 percent of all TBA sales were made by service stations, but by 1970 it was down to 30 percent. Significant inroads into service-station TBA sales have been made by discount stores, mass merchandisers, and specialty stores.[33] The low-volume service station is simply not an efficient or particularly effective competitor in an area of sales which one would expect it to dominate. However, if the oil companies are successful in their efforts to replace the traditional service station with the super station they may be able to hold their position in the TBA field and possibly improve it.

As a consequence of the difficulty that traditional service stations have had in maintaining their nonautomotive sales, there has been considerable experimentation in recent years with a variety of tie-in propositions. One of the more successful approaches has been the car-wash tie-in. Trailer and car rentals through service stations are increasing in frequency. Some majors are experimenting with tying together gasoline and convenience groceries. While some of these approaches are catching on, it is doubtful whether they will be able to correct some of the more fundamental problems of the major-brand service station business.

Method of Operating Service Stations

The nonprice competitive strategy of the major oil companies has been described and analyzed. The dominant method of competition was shown to be one based upon:

1. sales of branded gasoline;
2. relatively high and uniform prices;
3. service stations selling a single brand;
4. large numbers of conveniently located stations;
5. tools to increase customer's commitment:
   - credit cards,
   - trading stamps,
   - short-run promotions;
6. increasingly elaborate physical facilities; and
7. TBA sales and minor repair work.

What remains to be discussed and analyzed is the method by which the major-brand service stations are operated and the means by which gasoline and other automotive products are distributed to the stations.

Use of Independent Dealers

BY OWNING and controlling most of the better service stations the majors are in a position, if they so choose, to operate their own service stations with company employees. Were the majors to operate their stations, they could maintain detailed control over the retail setting in which their gasoline is sold. However, while company-operated stations were at one time the prevalent practice of the major-brand oil companies, this is no longer the case. For example, during 1968 Gulf operated less than 150 of its 25,000 service stations—less than 1 percent, and others operate even fewer of their own stations.

Instead of directly operating their own stations, the major oil companies (or their agents, the jobbers) rely primarily on "independent dealers" to run their service stations. The normal pattern is for a dealer to operate only a single station. Thus the prevailing arrangement by which major-brand gasoline is sold is by small "independent dealers" who operate a single station for its huge landlord-supplier, the major oil companies. This arrangement, by which the major oil companies are both the landlord and supplier of small independent operators, is rather unusual and its rationale will be explored.

A series of events during the 1930's resulted in the massive withdrawal of most major oil companies from the direct salaried operation of service stations. The impetus to move in this direction was in part associated with state legislation against the chain stores. The abandonment of salaried operation is often referred to as the "Iowa Plan" since this was the first state to pass such legislation. The plan was adopted by most of the major oil companies in Iowa to avoid a severe chain-store tax imposed by that state in 1935. However, the reasons which led major oil companies to lease their service stations to dealers are more complex than the avoidance of chain-store taxes.[34]

Many companies had begun to use dealers before 1935. Apparently, the Iowa Plan was widely adopted since companies found that it was more profitable for them to lease their stations to dealers than it was for them to operate them with salaried employees under the depressed economic conditions of the 1930's. Contributing to this unsatisfactory profit picture was the low average gallonage per station due primarily to the excessive number of retail outlets that the drive for market coverage had produced. Also, the small company-operated stations seemed to suffer from the pricing and operating inflexibility characteristic of this method of manning stations. Given a combination of low average gallonage and higher cost per gallon pumped, employee-operated stations often earned very low or negative returns.[35]

When a station was leased to one of the company's former employees there was often an increase in gasoline gallonage. In part this resulted from a better response to localized price cutting and in part to the greater personal initiative shown by dealers who were now working for themselves. In addition, dealer operations were more economical since they were free from chain-store taxes and Social Security contributions. Furthermore, dealers were frequently willing to work longer and harder for less and were often able to exploit the free- or low-cost labor of family members.

An additional reason the major oil companies operate with dealers is to reduce the likelihood of costly unionization of station employees. Since the major oil companies primarily operate through "technically independent" dealers, usually the hundreds of thousands of dealers are the bargaining units and not the major oil company. It is very difficult and expensive to try to organize the dealers' employees and most attempts to unionize them have not been particularly successful. Furthermore, the unions have found that the small operating profit of the major-brand stations is a further detriment to their wage demands—the dealers are squeezed too tight to be able to grant the level of wages bargained by the unions.[36] Thus, the dealers have become effective buffers to widespread unionization attempts and the major oil companies are able to operate their stations with a much lower level of wages and fringe benefits than would be the case if they were run with salaried employees.

The dealer method of operating also acts as a political buffer for the major oil companies. There is something very important in our cultural heritage about the small businessman striving to build a business that we want to protect. Coming between the huge oil companies and big government are some 200,000 so-called independent dealers. The major oil companies derive some favorable association by utilizing seemingly independent dealers to operate their stations. The dealers' problems with the major oil companies also divert some attention from other sensitive areas of the operations of the companies. Finally, there is something in the idea that if one attacks the major oil companies one is also attacking thousands of independent businessmen.

Method of Controlling Independent Dealers

IT SEEMS to be sufficient from a control standpoint for the major oil companies to own most of the better locations and to use independent dealers to operate their facilities. Since the dealer does not own the location his bargaining power is practically nil and his behavior can be directed in a large part by the actions of the major oil companies, which use a variety of techniques to bring about conformance to the operating procedures they feel to be important to their own performance.

The key aspect of the major oil companies' control over their dealers' operations is the short-term lease contract given dealers. Throughout the 1960's most dealers operated under one-year lease contracts with thirty-day cancellation clauses for infractions of the contract. The short-term nature of the contract puts dealers in the position where they realize that unless they generally operate in the manner prescribed by the major oil companies they run considerable risk of not having their contracts renewed when they expire. This gives the major oil companies considerable power since dealers often have big personal investments in time and effort building their businesses and also a financial investment in parts and inventory which are primarily of value to a going business. Recently some oil companies began experimenting with three-year dealer contracts that will supposedly give proven dealers more security. However, the national dealer association charges that this is a hollow gesture since thirty-day cancellation clauses are still part of these contracts.[37]

Major-brand dealers are supervised by company sales representatives. As first-line supervisors it is their responsibility to see that the dealers operate in the manner prescribed by the oil companies. The techniques they use range from friendly discussions with cooperative dealers to taking a less cooperative dealer aside and laying it on the line. There are several dimensions of the dealer's operation that are closely supervised.

The major oil companies do not solely rely upon personal communication between the dealer and the sales representative to closely control dealer operations. Some companies set the pattern of operation for their dealers by actually competing with them through strategically located company operations. In such cases it becomes a matter of not only "doing as I suggest," but "doing as I do." Two companies that use relatively large numbers of company-operated stations that frequently set the pattern for the dealer operations are Standard Oil of California and Sohio.

Another technique that has been used to control dealers' operations is to make dealers work on commission. The commission plan, "C" plan, is a hybrid program. It is similar to a company-operated station in that the oil company sets most of the operating policy and pays many of the bills. On the other hand, the "C" plan parallels the dealer program since the commission dealer's revenues are 1 percent of sales (not fixed) and the dealer's effort has a direct bearing on his earnings from the station. During the first half of the 1960's Shell had approximately 10 percent of its stations operating under the "C" plan and ten of the twenty-one largest oil companies had similar programs. As Shell said, these were "pacesetter" stations and the way they were operated influenced the behavior of the dealer-operated stations.[38]

Advertising is another technique used by the major oil companies to influence the behavior of their service-station dealers. When a major oil company runs a promotion of glassware, soda pop, or games, they seek widespread dealer acceptance of the promotion to maximize its effectiveness. Dealers frequently find themselves pressured to accept a promotion as a result of the extensive advertising supporting it. Whatever the promotion it is common for the advertising ballyhoo to state that the promotional offer is available at participating dealers. Dealers find that if they do not participate they spend much time explaining to customers the reason why.[39] Futhermore, such promotions require that dealers share the cost of the promotion which can run several hundred dollars a month.

Through the use of a wide variety of price-protection programs the major oil companies effectively determine the prices at which their dealers sell gasoline (see detailed discussion in Chapter 7, pp. 196–98). The gross margin of most dealers is too small to permit them to reduce prices on their own. If large numbers of dealers are going to lower prices, their suppliers have to reduce the price they charge their dealers for gasoline—their product cost. Where suppliers feel that it is to their benefit to encourage certain of these dealers to lower their prices, they reduce the dealers' product cost by granting price protection. The typical price-protection program results in the major oil company reducing dealer cost 0.7¢ a gallon for every 1.0¢ reduction in the suggested retail price. This means that the dealers' margin is reduced 0.3¢ per gallon for every 1.0¢ reduction in the suggested retail price to the point where a minimum margin is guaranteed for the dealer (the stop out). The theory behind this practice is that the dealer should not profit as a result of reduced prices, but rather be anxious to have prices restored to the normal level where he receives a full margin of profit. While oil companies are prohibited from setting dealers' prices, there are frequent indications that the majors have withheld price assistance from stations not reducing prices to the suggested level.

While price protection is the principal method by which the major oil companies control the prices at which dealers sell gasoline, it is not the only method that the majors employ to determine dealer pricing policy. Another major operating cost of the dealers which the major oil companies can manipulate is the rent dealers pay for their stations. By a variety of rental-concession programs the major oil companies can induce certain of their dealers to cut prices when desirable. The normal rental plan for major-brand stations is so many cents per gallon —frequently 1.5¢ to 2.0¢ per gallon.[40] Where a major oil company finds it advantageous for some reason to encourage certain of its dealers to reduce prices it waives rental payments, or sets low maximum gallonage figures on which it will collect rent—say 20,000 to 30,000 gallons a month. Those dealers receiving special rent concessions are put in a position in which it makes good business sense to reduce their prices below the normal market level and sell gasoline on a price-discounted, high-volume basis. Detroit is a market where special rental concessions have been particularly prevalent in certain parts of the city by companies such as Mobil and American Oil and have given rise to a large number of so-called maverick major-brand dealers. There are many examples in Detroit where certain major oil companies have taken low-gallon stations and turned them into large-volume producers by giving select dealers special rental concessions enabling them to cut price.[41]

Another technique that has been tried by the major oil companies in attempting to control the pricing of the dealers is fair trading. By fair trading their products the major oil companies establish a level of prices below which dealers are not permitted to sell their gasoline. Fair trading was instituted in five states—New Jersey, Pennsylvania, Connecticut, Rhode Island, and Massachusetts—in the early part of 1956 to stop some major-brand dealers from upsetting markets by price cutting. However, the only state in which it lasted any time at all was New Jersey.[42] Fair trading came to an end in New Jersey in 1969 when Esso abandoned the practice.

Controlled Aspects of Dealer Operations

THE CONTROLS that the major oil companies try to maintain over their dealer operations are normally those that they believe are in their best interests. To the extent that the desired practices of the oil companies and dealers are compatible there is little conflict. However, when the interest of the two are incompatible, or are viewed by some to be so, there is likely to be open disagreement over policy between the majors and some of their dealers.

The self-interest of the major oil companies as contrasted to that of their dealers breeds some conflict. The sales representatives of the major oil companies are often volume oriented since the fundamental purpose of the service station is as an outlet for products refined from crude oil. On the other hand, the dealer may personally find it more profitable, or desirable, to operate at lower-volume levels than desired by the major oil company.

While by no means the only factor of importance, the pricing practice of the dealer is of great concern to the oil company. During the extended gasoline hearings of the Federal Trade Commission of 1965 Commissioner Jones said, after hearing from witnesses of opposing viewpoints, that she gained the impression that if a dealer increased his price, or decreased his price, the major oil company would disapprove.[43] Commissioner Jones's observation comes fairly close to presenting the situation as it really is. For example, if Dealer A increases his price above the reference price, the volume of his supplier—Major Oil Company A—will decrease, which is a condition that Major Oil Company A will oppose (see Table 2-9—Competitive States 1 and 2). At the other extreme, if Dealer A decreases his price below reference level, Major Oil Company A's volume will increase. However, since competitive retaliation is likely, the gain will only be in the short run and the consequences may well be a costly price war (see Competitive

TABLE 2-9

Consequences of Dealer Pricing Practices on Major Oil Company Position

		Dealer A			Major Oil Company A			Dealer B		
Competitive State	Description of Price Condition	Price	Gasoline Volume	Profit	Price	Gasoline Volume	Profit	Price	Gasoline Volume	Profit
1	Both dealers selling at reference price.	O	O	O	O	O	O	O	O	O
2	Dealer A increases price above reference.	+	–	?	O	–	–	O	+	+
3A	Dealer A reduces price below reference.	–	+	?	O	+	+	O	–	–
3B	Dealer B reduces price to meet dealer A's price.	–	O	–	O	O	O	–	O	–
3C	Major oil companies give financial assistance to dealers resulting in price war.	–	O	–	–	O	–	–	O	–

O no change   + increase   – decrease

States 3A to 3C). Thus, the major oil companies are often opposed to dealers cutting prices. As previously discussed, the maverick major-brand dealers in Detroit (i.e., those selling below the reference price) in the second part of the 1960's were one of the major reasons for the severe price wars Detroit experienced during 1969 and 1970. The pricing practice most majors encourage is that their dealers price at, or within, one cent of the reference price and, furthermore, that dealers not post their price when markets are at or near the reference-price level. However, during price wars the majors encourage their dealers to become price aggressive and post their price on street signs.

A dealer's hours of operation are of prime importance to the major oil company and are sometimes stipulated in the dealer's lease contract. However, often there is difference of opinion over staying open early and late hours, or maintaining a twenty-four-hour-a-day operation. Dealers frequently feel that such hours won't pay, and this becomes a point of contention with the sales representative of the major oil company. When sales representatives and dealers agree about the cost of off-hour operation, the major oil companies at times give rental concessions to defray the cost. In other cases, hours of operation become the bases for lease cancellation. The following is a letter of cancellation supposedly sent by Shell Oil Company to a dealer in Detroit. The letter refers to the fifteen-day cancellation clause that allows Shell to put a dealer out of a station after a two-week notice.

REGISTERED MAIL
RETURN RECEIPT REQUESTED

Detroit, Michigan*

Reference is made (a) to our Lease with you as Lessee, dated October 18, 1967, covering the service station premises located at , in Detroit, Michigan, and (b) to our Dealer Agreement dated October 13, 1967, covering your purchase of "Shell" motor vehicle fuels and automotive lubricants for resale at that service station.

Your attention is specifically directed to Article 5 of said Lease, which states in part . . . "station shall be kept open for the sale of such products by Lessee at least 24 hours each day, excepting none"; and to Article 9 of said Lease which states in part . . . "In the event Lessee defaults in the performance or observance of any covenant or condition of this Lease, and fails to remedy same within 15 days after Shell gives Lessee notice thereof;—Shell may at its option and without notice, terminate this Lease."

This is to notify you that your failure to keep the agreed and specified hours of operation for the service station premises is a direct violation of the Lease and must be corrected by August 28, 1968.

In the event you fail to return to 24 hour operation by August 28, 1968, this is to notify you that we hereby terminate both the Lease and the Dealer Agreement effective on August 28, 1968. This is also our demand that you surrender possession of the premises (as defined in the Lease) to us on that date.

Very truly yours,
R. W. Lowe
District Manager

Another volume-building practice the major oil companies encourage is for their dealers to give trading stamps. Similar to long hours, trading stamps are frequently an expensive proposition for the dealers. Stamps for a low-volume station may cost $200 a month and for a high-volume station over $500 a month. A study carried out for

* *Prepared statement of Charles J. Hall before Senate Antitrust Subcommittee, July 14, 1970.*

the American Petroleum Institute stated that "the service station has difficulty offsetting the cost of stamps" and that "the pricing requirements which prevail in gasoline retailing (nearly uniform prices) make it difficult for service stations to break even on stamp plans." [44] It should be clear why dealers frequently resist the efforts of company sales representatives encouraging them to give trading stamps. Nonetheless, there is considerable evidence that dealers are frequently pressured to do so. Similarly, dealers have testified to arm twisting by the major oil companies to force them to participate in games of chance. The charges of coercion of dealers by oil companies to participate in their games was frequently heard during the Small Business Subcommittee's and Federal Trade Commission's hearings on games in 1968 and 1969.[45]

Another practice of the major oil companies that results in conflict is the dealer's purchases of tires, batteries, and accessories. While TBA sales do not have any direct relationship to gasoline volume, they do represent another source of revenue to the oil companies. As a result the major oil companies try to convince their exclusive dealers that they should purchase their TBA merchandise from the oil company. However, the Supreme Court ruling in the 1965 Atlantic Oil Company case outlawed the practice of an oil company supplier forcing its dealers to carry an exclusive TBA line—a practice for which it received a 7.5 percent override. A year later Texaco's and Shell's policy of selling dealers their private-brand TBA on an exclusive basis was also ruled to be illegal. While there is no doubt about exclusive dealing being illegal this has not prevented oil companies from exerting considerable effort to sell their line of TBA to their dealers. For example, a regional manager of American Oil Company said, "You can bet your life my men are out pushing our own TBA line to the dealers. No one ever took that right away from us. But that doesn't mean that we are going to club them over the head if they don't." John Huemmrich, executive director of the Congress of Petroleum Retailers, disagrees: "The methods may be a little more subtle, but they are just as forceful." [46] Alleged case histories of the ends to which certain majors go to sell their dealers TBA are sometimes quite bizarre.[47]

Other elements of the retail setting that the major oil companies are interested in controlling are associated with station image. These factors include station appearance, cleanliness, and customer services. While these are important factors in station operation, they seem to also be used as a justification for lease cancellations.

A symptom of problems in the service-station business is high dealer turnover. An American Petroleum Institute study during 1968 indicated an annual turnover of dealers of 23.9 percent. The Shell Oil Company's record during that year was 26 percent. In markets where price-war conditions are bad, turnover may hit 40 percent per year. The turnover rate of the gasoline industry is much higher than that of other industries employing independent dealers.

The plight of a good number of major oil company dealers is illustratively presented by the following:

Federal Trade Commission's *Report on Anticompetitive Practices in the Marketing of Gasoline*, June 30, 1967,

> "As a result of the marketing practices on the part of the suppliers (major oil companies), the retail dealer's position is largely that of an economic serf rather than that of an independent businessman;[48]

Wall Street Journal, October 1969,

> Service Station Men Say Big Oil Companies Keep Them in Bondage. . . . The big oil companies, as both supplier and landlord of the theoretically independent retailers, dictate prices, hours, operating procedures and products to be sold;[49]

National Director, Congress of Petroleum Retailers,

> One major oil company is calling in its dealers for face-to-face conferences and each dealer is being asked to sign a pledge to (1) Stay competitive on gasoline; (2) Give trading stamps; (3) Operate 24-hours a day; (4) Keep minimum house-keeping standards;[50]

U.S. Senator Thomas J. Dodd,

> A gasoline dealer is threatened by the demands of the very corporation that is responsible for his being in business, and . . . by increasing criminal activity, which so often victimizes him and his business.[51]

Hybrid and Direct Methods of Operating Stations

WHILE THE major oil companies rely primarily on theoretically independent dealers, they do to varying degrees employ other methods for operating a relatively small portion of their service stations. There are two primary arrangements that permit the majors more closely to control their service-station operations. One of these is a hybrid commission-consignment plan where the dealer comes quite close to being a company employee while still being classified as an independent businessman. Approximately 35 percent of Sohio's stations are operated in this manner while less than 1 percent of American's and Gulf's stations are commission operated. In essence what the commission-consignment dealer does is to operate the service station according to the dictates of the oil company; for this he receives a commission on sales from which he pays his employees and draws his own salary. The commission dealer differs from the typical dealer in that he has even less discretion over how he will operate the service station, and he differs from a company employee in that his earnings are variable and uncertain and he does not receive many of the employee's benefits. Historically the commission-consignment dealer has been used by the major oil company where it has been important for the major to gain control of prices at which dealers sell. The commission stations have played a role in controlling the prices of the other major-brand dealers and price marketers, especially in areas torn by price wars. Legal questions concerning this arrangement, such as the setting of dealer price by the major oil company and the legal classification of a commission dealer as an independent businessman rather than a company employee, have contributed to its limited use. In addition, some companies have used the commission plan for opening new stations.

The most straightforward manner the major oil companies can use to gain control of their service-station operation is to run them with company employees. However, there are several reasons why in general the majors operate relatively few of their own stations. These include the high cost of operation; the chance of unionization; the loss of personal initiative in station operation; and the possible loss of a political ally. The companies with the larger percentage of employee-operated stations include Standard Oil of California and Standard Oil of Ohio with close to 15 percent of their stations being directly operated. At the other extreme are American, Shell, and Mobil with less than 1 percent employee-operated stations. In the cases of Standard Oil of California and Sohio the company-operated stations are the bigger-volume stations where direct operations are economically feasible. These stations set the standard of operation sought for the dealer-run stations. Such stations are also a training ground for company employees and serve as reservoirs from which dealers come. While companies at times may employ commission dealers or salary operate their stations, the dominant and prevailing method of operating service stations is the dealer arrangement.

Selling and Distributing Gasoline to Stations

Gasoline jobbers and direct company operations are the two principal methods for selling and distributing gasoline to service stations. The jobbers are independent businessmen who purchase gasoline from the major oil companies and resell it to branded service-station dealers (as well as to commercial accounts). While jobbers sell branded gasoline to some independently owned service stations, a good portion, if not a majority of the gasoline, is sold to service-station dealers leasing stations owned or controlled by the jobber. Actually it is the owned and controlled service-station gasoline gallonage that makes the jobber important to the oil company. To sell gasoline to service stations and commercial customers the jobber normally owns some storage capacity for gasoline and other oil products, maintains an office from which the operation is directed, and owns delivery equipment ranging in size from small tank-wagon delivery trucks to transport trailers to serve volume accounts.

Direct company operations closely parallel those of the jobber with the primary difference being that the company owns the operation rather than the jobber. Sometimes the major oil company will use "commission agents" to deliver gasoline and other oil products to customers designated by it. Under such arrangements the oil company actually sells the accounts, owns the storage facility and retail outlets, does the billing, and extends credit.

Some companies rely heavily on jobbers while others give primary emphasis to direct operations for selling and distributing gasoline. For example, Sohio uses direct company operations to sell and distribute gasoline to practically all of its branded outlets in Ohio. Sohio's reliance on a direct operation is associated with its concentrated market and deep market penetration; it has over a 30-percent share of the Ohio market. In contrast, Phillips, which places considerable emphasis on jobbers, operates in all fifty states with relatively thin representation in most markets.

While some companies are committed to one approach as opposed to another, the most prevalent practice is for oil companies to follow a mixed approach—in some areas direct, in other areas jobbers, and in some areas both jobbers and direct. The tendency during the 1950's and 1960's has been for companies increasingly to follow the direct method in cities with populations of 50,000 or more. In the smaller markets, and particularly those with populations under 50,000, the major oil companies generally depend upon jobbers for selling and distributing gasoline.[52] Some of the majors also use jobbers in major markets where their representation is relatively weak and others use jobbers in major markets to supplement their direct operations.

Direct Operations

THE INCREASING reliance on direct operations for selling and distributing gasoline in a larger perspective is part of the process of vertical integration. The selling and distributing of gasoline represents one of the vital links in the series of activities that takes place from the time and place that crude oil is found and pumped out of the ground until the customer buys branded gasoline at a given service station.

Some of the specific reasons why jobbers are gradually being frozen out of the bigger gasoline markets includes investment requirements beyond the financial ability of jobbers; desire for secure market representation; elimination of problems created by jobbers; and greater consistency in the appearance of retail outlets. First, the investment requirements to put in service stations in metropolitan markets have been growing astronomically. Today's metropolitan stations cost from $150,000 to $250,000. Such station costs are simply beyond the financial ability of the vast majority of jobbers. For most jobbers to build only a small number of new stations the major oil company would have to lend its financial support so the oil companies might as well operate them directly. Furthermore, the returns from these investments are often poor from the jobbers' standpoint, but not to the integrated company that has big earnings from crude oil for which it needs outlets.

Secure market representation is another reason for direct operations in large markets. If the oil company is simply a supplier of a jobber, the jobber may be lost by a more attractive offer of a competitor because of the similarity of gasoline and the relative ease of switching brands. As long as major-brand competitors honor the status quo there is little problem with using jobbers. Some stability in jobber-supplier relationships is gained from fear of competitive retaliation for raiding another company's jobbers. However, outsiders have not honored the live-and-let-live practice of directly competing companies. The classic example of the outsider raiding other companies' jobbers involves Phillips' "Drive to Maine."

Phillips' invasion of the Southeast took three years. By the end of 1957 Phillips had 151 jobbers in the six-state area selling through 2,774 retail outlets. Eighty percent of the jobbers were in business selling some other brand before Phillips bought its way in with better deals for the jobbers.[53] During 1959 and 1960 Phillips contracted 32 jobbers in Ohio with 200 outlets and had commitments to build another 100 stations.[54] Beginning in 1961 Phillips launched its march into the mid-Atlantic area. The goal in this area was 1,000 stations in three years. An example of what took place in the wake of Phillips' invasion of the mid-Atlantic area was the switch of Aero Oil Company—one of Atlantic's largest jobbers—to Phillips, which involved nearly 200 service stations in nine Maryland and Pennsylvania counties.[55]

Similarly, other companies have expanded into new areas or deepened their penetration of given markets by the raiding of competitors' jobbers. For example, Sun Oil Company in 1970 bought out the Mobil jobber in St. Louis with approximately 80 stations and a large Gulf jobber in Rockford, Illinois, with more than 200 stations. This practice has very definitely contributed to the major oil companies' drive for direct operation or control over jobbers, as will be discussed.

Major oil companies are inclined to engage in direct wholesaling operations in big markets to eliminate problems created by some of their jobbers. Frequently an aggressive jobber will upset a market by engaging in a variety of price-cutting practices which end up costing his supplier and all others a great deal of money. What happens is that the aggressive jobber induces his dealers to cut prices to build volume. Eventually competitive dealers and jobbers react by cutting their price and ultimately all the major oil companies find it necessary to give price protection to their dealers. Unless the aggressive jobber changes his behavior the market may be in for a series of price wars. Very possibly the major oil company with the aggressive jobber will find itself in a position of having to buy out the jobber at a premium price. As an example, there are indications that certain major-brand jobbers have been primary contributing factors to problems in Detroit. Evidence indicates that Gulf had a couple of aggressive jobbers in Detroit in the latter 1960's and found it advantageous to purchase the jobberships. Similarly, testimony given at the Senate Antitrust Subcommittee's hearing on pricing problems in Detroit in July 1970 suggested that an American Oil jobber might be a major factor contributing to the market's instability. Subsequently, American Oil cancelled its long supply relationship with the jobber, Citrin Oil Company.

A final reason majors often prefer direct operation to jobbers in large markets is to maintain greater consistency in the appearance of the service stations selling their brands. The appearance of the jobber stations is typically quite mixed, ranging from some new, modern stations to some that are "old dog." While the majors may desire to replace the older stations with new and modern ones, they can only do this to the extent that they can convince the jobber to act. This condition has also been used by majors as a premise for entering markets in direct competition with their jobbers.

Use of Jobbers

WHILE SOME JOBBERS have significant positions in the bigger markets, this is not where their strength lies. The approximate 20 percent of the nation's gasoline that jobbers sell is heavily concentrated in the smaller communities where jobbers seem to have certain natural advantages. In the lesser developed markets, the jobbers may have an economic edge over the large companies with their complex organizational structures. These are also the markets that are stagnant or not expanding as fast as those in the bigger cities and there is less of a need for investment in new outlets. Furthermore, in the smaller communities pressures to upgrade the appearance of physical facilities are not as strong as they are in many of the big cities. In addition, when new service stations are built, the cost of land and improvements is considerably less. Another reason for the jobber's strength in the smaller markets is that the recognition of who the man is, and that he is part of the community, is much more important there than in urban centers.

It is also less risky for the major oil companies to rely more heavily on jobbers in the smaller markets than in the larger ones. To lose a jobber selling gasoline to twenty relatively low-volume stations is quite a different matter than to lose a jobber in a big market with 100 high-volume stations. The normal attrition of smaller jobbers and the development of new ones tend to cancel each other out. However, when a large jobber is lost it is difficult to replace such business.

The oil companies reduce the likelihood of a jobber switching by writing restrictive contracts with their jobbers when it is at all possible; this, of course, depends upon the strength of the jobber. For example, many jobbers are unable to get long-term financing (fifteen years) on their own to modernize and to expand. Thus, they turn to their suppliers for financial assistance. If the supplier and jobber agree to build a new station, it frequently will be leased to the major oil company for fifteen years. With this insurance of revenue, the supplier's bank or some other financial institution will make the loan to the jobber. However, the jobber is "on the hook" to the supplier for fifteen years and possibly longer if the supplier has options to extend the lease. When a jobber has a number of his stations so tied up it becomes impractical for him to switch suppliers for he would then lose the volume from the newer and more modern stations that he has been building for a number of years.[56]

Oil companies also control their jobbers through the supply contract. Frequently, the supply contract includes a restrictive first refusal option which states that the present supplier has a ninety-day period to accept the terms and conditions for which a jobber is willing to sell his operation. What in essence this option requires is that a jobber make known to his present supplier any offer that a competitive oil company makes for the jobbership. This very much discourages another supplier from extending an offer to purchase a jobber for it is recognized that the offer will have to be submitted to the present supplier. As a result many jobbers find today that if they wish to sell out the only real potential buyer that they have is their existing supplier.[57]

Cost of Marketing Major-brand Gasoline

It should be readily apparent that the nonprice competitive style of the major oil companies is a very costly method of selling gasoline to the public. A characteristic of this nonprice method of competition is that through time more and more cost gets built into this method of marketing gasoline. In the short run, oil companies can temporarily gain a competitive edge by innovatively employing nonprice competitive tools. However, as more competitors adopt these practices, other more enticing tools have to be developed. For example, following World War II the emphasis was on brand-name advertising of gasoline sold through large numbers of low-volume, conveniently located stations. During the 1950's the oil companies increasingly employed credit cards and trading stamps to tie customers to their branded stations. The 1960's saw the sales of gasoline on oil-company credit cards mushroom. During the second half of the 1960's, games became the rage. About the same time the "elaboration" of service-station facilities started and it continues through today. Most of these nonprice competitive tools have gradually become institutionalized as necessary for majors to sell successfully their branded gasoline and they have added substantially to the majors' cost of selling gasoline.[58]

Some idea of the relative cost of the nonprice competitive style of the major can be obtained from the approximate cost schedule for marketing major-brand gasoline presented in Table 2-10. St. Louis is

TABLE 2-10

Approximate Cost of Marketing Major-brand Regular Gasoline in St. Louis, Fall 1969

Price per gallon of regular:			35.9¢[1]	
State and federal tax:			9.0	
Retail price minus tax:			26.9¢	100.0%
Dealer margin (tank wagon 18.2¢):	8.0¢	32.4%		
Jobber margin:	3.75	13.9		
Jobber delivery allowance:	.45	1.6	12.9	47.9
Branded price from supplier:			14.0¢	52.1%
Branding margin:			1.5	5.6
Unbranded price delivered:			12.5¢	46.5%

1. *Platt's Oilgram Price Service* throughout the fall of 1969 indicated the service station price to be 34.9¢ for major-brand regular gasoline while at the time the survey was done the prevailing major-brand price was 35.9¢.

used as an example since it was one of the three markets studied and cost figures were more readily available for this market than for the other two. The prevailing retail price of regular-grade gasoline excluding tax was 26.9¢. The dealer margin, jobber margin, and delivery allowance combined amounted to 12.9¢ per gallon, 47.9 percent of the retail price. The branding margin (the difference between the wholesale price of branded and unbranded gasoline) is 1.50¢, 5.6 percent of the retail price. In total the approximate marketing cost for major-brand gasoline in St. Louis was approximately 14.4¢ per gallon of regular-grade gasoline (26.9¢ − 12.5 = 14.4¢), or approximately 53.5 percent of the retail price of gasoline before taxes.

The price marketers' strategy developed in the next chapter makes sense and only exists as a result of the relatively high cost of the non-price competitive strategy of the major oil companies for marketing their branded gasoline. In essence, what the price marketers have done is move into the void created by the homogeneous strategy of the majors. The price marketers' appeal is to that segment of the population that prefers lower price rather than the more costly major-brand gasoline.

PLATE 2-1 Functional Porcelain Box-type Station

PLATE 2-2 Increased Size of Traditional-type Station

PLATE 2-3 Shell's Ranch Design Station

PLATE 3-1 Early Tank Car Station of Martin Oil Company

PLATE 3-2 First Self-service Station of Urich Oil Company

PLATE 3-3 Minimum Investment Station of Widely Dispersed Thoni Operation

PLATE 3-4 Limited Service Station—Save Way Operation

PLATE 3-5 Net Operator—A Mary Hudson Station

PLATE 3-6 Holiday Merchandise Station

PLATE 8-1 Station Being Converted From a Private Brand to Gulf

3 / *Marketing Style of Price Discounters*

Trackside Stations

The modern-day price marketer had its origins in the "trackside station" which emerged during the middle 1920's and grew during the grim days of the 1930's. The tracksiders located along railroad spurs and pioneered the practice of selling price-discounted gasoline for several cents less than the major-brand stations. This purchase proposition had broad appeal during the depression years.

The tracksiders were able to sell gasoline for less because of their operating efficiencies, their lower-cost gasoline, and their high station volume. Operating costs were reduced by partially or totally eliminating steps in the traditional methods of gasoline distribution. During the 1920's and 1930's, major-brand gasoline typically moved from refineries in railroad tank cars (in 8,000 gallon quantities) to smaller local storage facilities called bulk stations. Finally it was distributed by small delivery trucks with capacities of 200 to 500 gallons (called tank wagons) to service stations, which bought a few hundred gallons at a time. The tracksiders eliminated much of the cost associated with local storage and small-volume deliveries to stations by building their stations beside railroad spurs. This allowed them to sell gasoline to the public on a semiwholesale basis. As the sales proposition matured and proved itself, the tracksiders built large above-ground storage tanks for gasoline (see Plate 3-1) which saved them demurrage on the delayed release of railroad equipment and gave them greater flexibility.[1]

In addition to operating efficiencies, the depressed economic conditions of the 1930's often put the tracksiders in a position to buy gasoline at low prices. Since they marketed under their own brand, they were not committed to pay the going wholesale price set by the supplier for branded products. Instead, the price marketers bought their gasoline for the lowest price quotation in the open or spot market. Since gasoline refineries typically operated far below capacity during the depression, gasoline was purchased at bargain prices. When tracksiders could make particularly good buys, they would fill up their large storage capacity with low-price products.

The excesses of the 1920's also created conditions favorable to the success of the high-volume, low-price strategy of the price competitors. During this period new service stations had been built with little regard to demand. The result was a high-cost inefficient distribution structure consisting of an excessive number of low-volume stations.[2] It was this relatively high-cost, low-volume gasoline structure which created the void into which the tracksiders moved with their high-volume, low-cost, low-price stations. Because of their volume, tracksiders could profitably handle gasoline for three cents per gallon in comparison to the vast majority of their competitors who required six cents to operate profitably.[3] While a good major-brand station might be selling 5,000 gallons per month, the tracksider would often sell five to ten times as much for two to five cents less.

The majors tried a variety of techniques to check the volume drain of the price marketers. Early in the thirties several majors introduced third-grade gasoline as fighting brands to compete with the price cutters. On the West coast the majors marketed secondary brands through separate outlets to meet the price cutter competition head on. Other majors like Sohio cut the price of their regular-grade gasoline to compete with the price marketers where they were being hurt.[4] Despite all efforts to stop tracksiders, the acceptance of the price marketers' sales proposition of private branded gasoline for less grew during the decade.

Several of the conditions that had favored the growth of the tracksiders in the 1930's disappeared during the war years, 1941 to 1945. The oversupply of the thirties turned into a shortage, gasoline rationing was introduced, and the price competitors were unable to buy gasoline in the spot market on favorable terms. In addition, the war produced a shortage of railroad tank cars which meant that gasoline increasingly had to be distributed by transport trucks. These conditions and others reduced the competitive edge of the price competitors and hurt their position. They remained essentially dormant during this period.

After the war, conditions were once again favorable for the growth of price competition; however, the era of the tracksiders was over.

Barges, pipelines, and especially large-volume transport trucks were replacing railroad tank cars as more efficient methods of distributing gasoline. Since transport efficiencies were now available on a much broader scale, the tracksiders' competitive edge was obviously diminished. In addition, the decentralized character of urban growth meant that trackside locations were often not easily accessible to the expanding gasoline market. While the tracksiding principle was to disappear, some of the original trackside operations such as Site, Martin, and Tankar, were about to play a major role in newer types of price competition.

Early Self-service

Modern-day price marketing has been importantly influenced by the dramatic growth of self-service gasoline marketing in the postwar readjustment period. Large-scale, self-service gasoline operations were first introduced in the Los Angeles basin on May 1, 1947, by George R. Urich of the Urich Oil Company. By the end of the year, Urich had a total of three giant self-servers and competitors had two similar operations.[5] From this beginning, self-service stations spread up and down the West Coast, into the near West, through several states in the central part of the country, and then on to the East Coast.[6]

Characteristics of Innovation

Self-service gasoline retailing, in which customers dispensed their own gasoline and then paid a cashier, represented a major departure from the traditional method of selling gasoline. This innovative approach was tied to the development of the automatic shutoff nozzle for dispensing gasoline. Back pressure builds in the gasoline tank of a car as it is being filled with gasoline. The automatic nozzle has a sensing device that measures this pressure and automatically shuts off when the tank is almost full. This technological development permitted the self-servers to argue effectively with many fire marshalls that there was little danger in allowing customers to serve themselves.[7]

A second major difference between self-service operations and the traditional marketing of branded gasoline was the character and layout of the station. The new stations were very large and spacious and located on major traffic arteries. The street frontage for a self-server often ranged from 150 to 400 feet and occupied up to an acre of land. The self-service stations impressed drivers with their broad and flat expanses broken only by several rows of pumps. Another distinctive feature of these stations was the large number of islands (six to ten) and of pumps (eighteen to twenty) placed perpendicular with the street.[8] This was in contrast to major-brand service stations, which at that time occupied corners with 40- to 75-foot frontage and had one or two islands and two to four pumps.

Speed of operation was another important characteristic of self-service stations. Maximum access to and utilization of gasoline pumps were achieved by making the gassing of a car the sole operation at the pumps. To speed up the gassing phase of the operation, and to add a flare, some operators used girls on roller skates who scurried from island to island reading meters and collecting money. Service areas were located away from the pumps where drivers could buy and add oil, wash their car windows with materials provided, or further service their own cars.[9]

Large signs at the stations proclaimed SERVE YOURSELF—SAVE 5¢ PER GALLON, clearly presenting one of the major elements of self-service gasoline retailing—selling gasoline for less. To sell for five cents less than most major outlets self-service operators had to buy gasoline at low prices and also had to cut operating costs. Drawing on the trackside idea, the self-service stations had large underground storage facilities (30,000 to 90,000 gallons), which enabled them to take full trailer load dumps (up to 5,000 gallons) resulting in a savings in distribution costs. In addition, they sold unbranded gasoline, which cost considerably less than branded gasoline. The self-service operators also had lower-operating costs. One major saving was labor costs since customers served themselves, and one or two men could do the work of many. Another important saving was the low cost per gallon of physical facilities. The fixed facility costs—often high in total—were low on a gallonage basis because of the large volume of gasoline being sold. Some of the romance associated with early self-service is described in a letter from an executive of Urich Oil Company which appears in Appendix A.

Volume Operations

THE VOLUME of gasoline sold by several of these self-service stations was staggering. According to major company estimates at least six of the seventeen stations in existence at the beginning of 1948 were selling 150,000 gallons of gasoline per month. Of the six, two were judged to be selling more than 400,000 gallons per month and one more than 500,000 gallons.[10]

Urich's first eighteen-pump station sold 450,000 gallons of gasoline per month within three months of its opening (see Plate 3-2). This was twice the total volume of all the gasoline being sold by Urich's twenty-five conventional stations. Within a year and one-half Urich had six self-service stations and two conventional stations that were selling 2,000,000 gallons of gasoline a month. This is in contrast to 250,000 gallons per month that had been sold by all the twenty-five conventional stations. John Rothschild of Rothschild Oil Company (now Powerine Oil Company) said that they pumped more gasoline through their three eighteen-pump stations in a single day than many neighborhood stations pumped in a month.

In contrast to the Urich and Rothschild self-service operation, some self-service stations were simply conversions of conventional stations. The largest operator of converted stations was Eagle Oil and Refining Company, which converted 150 of their 600 Golden Eagle stations to self-service. The Golden Eagle experience showed that the conversion of a conventional station to self-service would quadruple its volume of business. A 10,000-gallon station would sell 40,000 gallons.[11]

Rise and Decline

As self-service grew, it became embroiled in controversy. Was it a "new method of merchandising with a different economic formula" or was it simply a "gimmick for cutting price"? The following statement by an executive of the Urich Oil Company typifies the self-servers' views on the subject.

> Mr. Jones [a conventional operator] may have 600 service stations pumping 3,000,000 gallons a month. His rents will run $175 to $200 each (around $120,000 a month). He needs 60 trucks, 60 men each, and 10 branch offices with personnel.
>
> Mr. Smith [the self-service operator] can do the same volume with ten self-service stations. Our rents are nothing compared to the conventionals. He will need six to eight trucks and drivers. One man can make seven trips in a day whereas one truck spends a day going

from station to station when serving the conventionals. Only one maintenance crew is needed and no branch offices are required.[12]

This executive went on to explain that the nickel given away to the customer was the savings coming from selling gasoline on a large-volume, wholesale basis as contrasted to the costly low-volume, convenience method of selling gasoline.

By late 1948, less than two years after Urich opened his first station, self-service stations were estimated to have over 5 percent of the gasoline business in the Los Angeles basin and more stations were planned. In many other areas where self-service had been introduced, it was making significant market inroads.

As the major oil companies, jobbers, and dealers saw their volume being drained away they took a number of steps to counter the growth of self-service operations. A major effort was exerted to outlaw self-service stations at the state, county, and city level. The American Petroleum Institute passed a resolution calling for a ban on self-servers. When supply contracts of large self-service operations expired they were not renewed, and major-brand dealers who converted to self-service found oil companies refusing to continue supply relationships. Beginning in early 1950 the major oil companies lowered their prices in those areas where self-service was becoming a significant market factor and a period of prolonged price wars followed. These three measures, along with others, which were employed in the fight against self-service operations, are developed in Appendix B. They are similar in effect to the techniques used to check the growth of price marketers in the early 1960's, which are detailed in Chapters 4 to 8.

By the end of 1951 self-service retailing of gasoline had been defeated in many areas where it was making significant market inroads. Several self-servers sold their operations and others began to give service and to increase their prices. While the first post-World War II era of self-service was over, it demonstrated the potential for mass merchandising of gasoline on an economy formula basis.

Conventional Price Marketers

While the West Coast self-service marketing of gasoline was a dramatic development, other types of price marketers were also making substantial inroads in many parts of the country. Certain changes were taking place in society that contributed to the increasing importance of these price marketers. In particular, the massive population shift to the suburbs and the increase in commuting by automobile created a key opportunity that was ready-made for the price marketer.

The extent to which price marketers have penetrated individual markets seems to be associated with two factors: first, the availability of an adequate supply of unbranded gasoline at competitive prices and, second, the availability of reasonably priced locations. Supply opportunities have tended to be better in markets near the major crude-oil producing and refinery areas of the Gulf and Mid-continent or in markets economically reached from these areas by barge, pipeline, and ocean tanker. In other regions independent refiners and small majors are important sources of supply. The general area in which favorable supply conditions have facilitated the growth of price marketers includes the South and Southwest, the Middle West, and the north-central parts of the country. The principal areas in which supply conditions have hampered private-brand growth includes the central and eastern Great Lakes area, the mid-Atlantic region, the New England states, and the West Coast. The growth of price marketers is related to the cost and availability of real estate. In certain major metropolitan markets the growth of price marketers has been inhibited by tight zoning and high land costs.

The price marketers who emerged in the latter 1940's and early 1950's were different in many respects from the trackside operators and other price marketers of the 1930's. No longer did they operate marginal, rundown, scrubby-looking stations that sold gasoline of questionable quality. Modern price marketers began to operate some of the most attractive stations in existence.[13] In fact, they are responsible for the era of the large multipump station. Of the estimated 800 multipump stations in existence in 1950 the vast majority were operated by price marketers.[14]

Unlike their counterpart of the thirties, the strategy of the price marketers in the fifties and sixties was to build large multipump stations on major traffic arteries that link the suburbs with the central city (see Figure 3-1). Stations located in such a manner were in a unique position to intercept drivers as they went to and from work. The high-traffic locations were essential for the modern price marketers' high-volume type operation. In some metropolitan areas where the suburbs

FIGURE 3-1

TRADITIONAL LOCATION OF PRICE MARKETERS IN URBAN AREAS

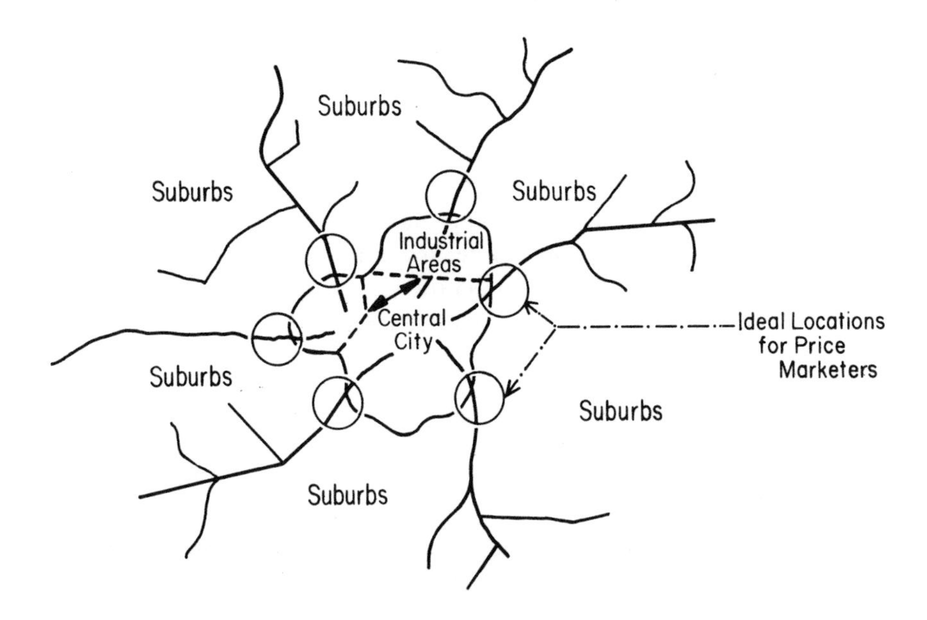

have grown very large, the price marketers have moved into the suburban areas in their search for good traffic locations. In other cases where expressways have bypassed once good locations, price marketers have relocated on major arteries feeding the expressways. In any event the traffic location is crucial.

Discount Philosophy

WHILE THE marketing mix of the modern price marketer is different in many ways from his counterpart of the thirties, price still plays a key role in his purchase proposition. The reason for the emphasis on price is clear. It is because the price-conscious segment of the market has generally been ignored in the marketing strategy of the major oil companies. A second reason the independents have followed a price strategy is that it minimizes the competitive disadvantage they face in making investments in retail locations. The price marketers conserve their financial resources by operating relatively few stations, which sell large gasoline volume.

The strategy of the modern price marketer is then basically one of low price, high volume, and low unit cost. There is nothing mystical about this formula. It is basically the same strategy followed by other mass merchandisers such as the supermarket and the general merchandise discount stores which have revolutionized their respective industries. When analyzing the modern price marketers' strategy, it is impossible to say which comes first—low prices, low unit cost, or high volume. These three elements complement and reinforce one another. For example, low unit costs make low price feasible. However, the actual cost per gallon depends upon the volume done—and the volume of business is a function of relative price levels among other things.

Comparison of Operating Margins

THERE ARE several ways in which independents effect savings which permit them to sell for lower prices. One saving is in the price they pay for their gasoline. The price marketers buy unbranded gasoline as compared with the branded gasoline purchased by major-brand retailers. Unbranded gasoline costs less than branded gasoline. The higher cost of major-branded gasolines is associated with the extra cost involved in brand-name advertising, credit-card sales, and sales-promotion programs—that is, from the costs of product differentiation and brand promotion. As a result of these savings the wholesale price of unbranded gasoline on a volume-purchase basis is one to two cents per gallon less than the price of branded gasoline during normal periods. The range in prices is associated with:

1. availability of supply of unbranded gasoline; and,
2. differences in the wholesale price of branded gasoline between markets.

The price spread between branded and unbranded price tends to narrow as one moves toward the East and West Coasts—particularly in the northern states in each area.

Besides a lower cost of gasoline, the price marketers often have lower fixed costs and especially lower operating costs. Real estate and improvements cost less. The price marketers typically stay away from the expensive corner locations. Also, since most price marketers do not offer repair service, their stations do not have bays and building costs are less. Savings in operating costs are associated with the relatively high station-volume of the price marketers. As a result of higher volume, labor productivity is high and labor costs per gallon are low. Similarly, fixed operating costs such as those for management, utilities, insurance, licenses, and taxes are spread over a large volume and are low on a per gallon basis. The savings that the price marketers have been able to effect give many of them a total operating cost including profit of approximately 8.0¢ per gallon. This is in contrast to the 11.0 to 11.5¢ (an approximate 4.0¢ jobber margin and a 7.0 to 7.5¢ dealer margin) that the major oil companies required at normal market prices to distribute their gasoline in 1970. When the operating margin of the price marketers of 8.0¢ per gallon is subtracted from that of the distribution cost of major-brand retailers of 11.0 to 11.5¢, the operational cost advantage of the price marketers is often in the neighborhood of 3.0 to 3.5¢ per gallon. Adding this saving to the saving from buying unbranded gasoline, 1.0 to 2.0¢ per gallon, the price marketers frequently have a cost advantage of from 4.0 to 5.5¢ per gallon. It is because of such savings that the price marketers are able to undersell the major retailers by several cents per gallon. The vice president of marketing for Continental Oil Company said in a speech during 1971 that in his markets the difference between the unbranded wholesale price of gasoline and the major-brand retail price ranged from 13.0 to 15.0¢ per gallon.[15] If one subtracts the unbranded operating margin of approximately 8.0¢ from the 13.0 to 15.0¢, the price marketers are frequently in a position to sell gasoline for 5.0 to 7.0¢ less than major-brand gasoline retailers.

Magnitude of Price Discounts

THE MAGNITUDE of the price differential between the price marketers and the major-brand station varies. During normal competitive periods the price differentials will be the greatest and they will narrow or will disappear during price wars. Market surveys carried out during the fall of 1969 give some idea of the differentials in three markets during relatively stable competitive periods. Washington, D.C., San Francisco, and St. Louis were selected as the cities to study to obtain geographic coverage of the East Coast, West Coast, and the Midwest (see Table 3-1). The price marketers were underselling the majors by

TABLE 3-1

DIFFERENCES IN THE PRICE OF GASOLINE OF MAJORS AND PRICE MARKETERS IN THREE GEOGRAPHICALLY DISPERSED AREAS

	Median Price Regular Gasoline	Median Price Ethyl Gasoline
Washington, D.C.		
Major brand:	35.9¢	39.9¢
Price marketer:	30.9	34.9
Difference:	5.0¢	5.0¢
San Francisco and West Bay Area, California		
Major brand:	36.9¢	40.9¢
Price marketer:	31.9	35.9
Difference:	5.0¢	5.0¢
Greater San Jose Area, California		
Major brand:	35.9¢	39.9¢
Price marketer:	30.9	34.9
Difference:	5.0¢	5.0¢
St. Louis, Missouri		
Major brand:	35.9¢	39.9¢
Price marketer:	32.9	34.9
Difference:	3.0¢	5.0¢

five cents per gallon in Washington, D.C. A similar spread between the price marketers and majors was found in San Francisco (divided into two parts: (1) San Francisco and the West Bay communities and (2) Greater San Jose area). In St. Louis the price differential between the price marketers and the major brands was three cents for regular and five cents for ethyl. In other markets where price wars are relatively frequent or where price marketers have a large share of the market, the price spread between private-brand and major-brand gasoline may settle out at an average of three cents per gallon (two cents on regular and four cents for premium). In such areas price marketers are likely to reduce their real selling price by giving high-value cash or premium redemption stamps.

Other Operating Characteristics

THE PRICE MARKETERS generally call attention to their lower gasoline prices by the use of price signs. In contrast, the major stations typically do not advertise price unless they are involved in a price war. For example, in the three cities surveyed, 68.7 percent of the price marketers posted their prices while only 15.9 percent of the major-brand stations posted their prices (see Table 3-2).

TABLE 3-2

SIGN POSTING OF PRICES BY MAJORS AND PRICE MARKETERS

	Major-brand Stations			Price Marketers		
	Number of Stations	Posting		Number of Stations	Posting	
Washington, D.C.	643	100	15.5%	54	44	81.5%
San Francisco and West Bay	384	88	22.9	41	38	92.7
St. Louis	431	44	10.2	116	63	54.3
	1,458	232	15.9%	211	145	68.7%

Furthermore, the methods of posting price by the price marketers and majors are generally quite different. The price marketers use large, permanent, and conspicuous price signs. In contrast, when the majors find it desirable to supplement their fundamentally nonprice strategy with more aggressive pricing, they frequently do so with temporary curbside "A" frame price signs. These signs can be placed quickly and removed as easily.

Another characteristic of price marketers is that a relatively large number of their stations are company operated rather than dealer operated. A survey of eighty-eight price marketers with more than 6,000 stations reveals that approximately half are operating their own stations with salaried employees.[16] There are several reasons for this emphasis on salaried operation by price marketers even in the face of higher labor costs. The operation of high-volume, low-price stations requires quick adjustment to competitive developments. This is difficult to accomplish under the dealer arrangement.

Another distinguishing feature of the price marketers' strategy is their hours of operation. The vast majority of major stations operate less than eighteen hours per day, while the reverse is true for price marketers. Sixty-nine percent of the price marketers' stations surveyed stay open from eighteen to twenty-four hours per day (see Table 3-3).

TABLE 3-3

HOURS OF OPERATION FOR EIGHTY-EIGHT PRICE-MARKETING CHAINS DURING 1966

Hours Per Day	Stations	
24	2,988	50%
18 to 23	1,164	19
14 to 17	1,520	25
Under 14	361	6
	6,033	100%

Source: Data compiled by Society of Independent Gasoline Marketers.

Price marketers typically do no repair work. As a result their buildings are smaller than those of the majors and involve less investment. The typical price marketer's station provides space enough for rest rooms, a place for attendants to get out of the weather and do their paper work, and a small storage area for oil and work supplies. The emphasis is on quick gasoline service through attractive facilities. Price marketers attempt to create an uncluttered streamlined appearance by eliminating parked cars waiting for repair work to be completed, by not parking a tow truck on the premises, and by staying out of the car and trailer rental business. Furthermore, by eliminating greasy repair work the appearance of price marketers' employees is often more appealing than that of attendants at major-brand stations.

In summary, some of the basic differences between modern price marketers and major-brand gasoline retailers are presented below:

Distinguishing Features of Price Marketers	*Distinguishing Features of Major-brand Gasoline Retailers*
Emphasis on lower price	Emphasis on advertised brands
High station gallonage	Lower station gallonage
Few centrally located stations	Many conveniently located stations
Spacious layout	Smaller properties
Sells primarily gasoline	Sells gasoline and TBA
Company-operated stations	Dealer-operated stations
Open long hours	Open shorter hours
Quick island service	Complete car service
No repair work	Perform minor repair service
Purchases unbranded gasoline	Purchases major-brand gasoline.

Different Styles of Operations

Even though price marketers have many characteristics in common there is still a great deal of difference among their operations. These differences mirror the strong personalities of the owners of the individual companies and the peculiarities of the markets in which they compete.

Deep Discounters

A rugged type of price marketer is one having a marketing philosophy of pricing one cent below everyone else. Often these operators take over abandoned stations where majors have chosen not to renew their leases. Since the stations are fully depreciated, the monthly rent may be less than half that paid by the major oil company abandoning it. In other cases an outright purchase of the station may be made for 25 to 50 percent of replacement cost. In many cases the only additional investment required may be three or four thousand dollars for pumps, paint, and an inexpensive price sign. In this manner the one-cent-under price marketers frequently have extremely low-cost facilities which aid them in implementing their pricing policy.

Price marketers who fall into this category include both those with far-flung multimarket operations and others who scatter their operation throughout major metropolitan areas. Examples of the widely scattered operators would include the Thoni brand stations (operating primarily in the South and Southeast)—see Plate 3-3—and the Hudson family with their Highway, Poor Boy, Tommy (operating from the Midwest to the East Coast). Those "deep discounters" who operate primarily throughout large metropolitan areas would include the Lerner brand stations in Los Angeles, the Texas Discount brand stations in St. Louis and Rufus Limpp stations in central and western Missouri. By running a "bare-bones" operation—low-cost facilities, one-man shifts, and no frills—they can and do sell for less than anyone else in many areas and markets.

There are many examples of such operators who have taken over former major-brand stations selling less than 10,000 to 15,000 gallons of gasoline per month. By following their low-cost and low-price method, the deep discounters frequently sell several times the gasoline previously dispensed by the station. Not surprisingly, such operators often become involved in price wars when competitors are unwilling to give them the price edge they demand. However, in many cases competitors choose to ignore them since price retaliation could bring down an entire market. They reason, "How much volume can they do through a handful of stations?"

Limited-service Operator

One of the off-shoots of the earlier self-service era is the limited-service type of operation. The attendants at these stations pump gasoline, add oil, and collect cash; but that is all they will do. Remote from the pumps are service areas where drivers can clean their own windshields, add water, and vacuum the inside of their cars.

There are relatively few limited-service type operations. Their scarcity is generally associated with the need for the limited-service operator to sell for less than other types of price marketers if he is to succeed. However, price competitors are frequently reluctant about allowing the limited-service operation to sell for less. Therefore, it takes either a very strong price marketer, or one who faces little direct price competition, to implement the limited-service strategy.

Price marketers following this strategy are companies such as Simas in San Francisco; Scot in Washington, D.C.; and Meadville with Save Way, Safe Way, and Giant stations in the East. While there are not many adherents to this competitive strategy, some of the world's largest stations are operated on the limited-service plan. Several of the Save Way, Safe Way, and Giant stations sell 150,000 to 250,000 gallons of gasoline per month—as much as ten times the national average. Their stations are often one-half block to a block long and are located on major arteries in the densely populated areas (see Plate 3-4). The appeal of these stations is that they are clean, open-looking, give quick service at the pump, and sell for several cents less than the major-brand competitor. Like other aggressive price marketers, they have scattered their operations geographically in order to reduce their vulnerability to regionally concentrated price wars. For example, Meadville's Giant and Save Way stations are located in thirteen cities, ranging from New York City to Chicago and from Boston to Washington, D.C.

Net Operators

NET MARKETERS are those who do not use stamps, premiums, or other techniques as ambiguous ways of extending their price discounts and who frequently take the position that they are justified in selling gasoline for less than those price marketers giving customers stamps and premiums. Like limited-service operations, net marketers often run into problems in implementing their strategy for other price marketers may be reluctant to allow them to sell for less, and the absolute amount of their price discount they take may be more than majors are willing to accept. As a result this type of operator is frequently involved in price wars. To reduce their vulnerability to price wars, the net operators normally disperse their stations over large geographic areas.

Examples of net marketers include Hudson Oil Company, Hudson of Delaware, Imperial Refineries, and Dixie Vim. Mary Hudson operates approximately 300 stations in thirty-four states. Her stations have conspicuous price signs with quick change numerals, wide street frontage with broad driveways, four to six islands spread far apart, uniformed employees, colorful billboards telling customers about Hudson and its products, and a small building for the employees (see Plate 3-5).

Stamps and Premium Givers

MANY of the price marketers either do not have the regional diversity necessary to implement successfully a net pricing policy or are located in areas where price marketers are more important and majors are more likely to react. These price operators have learned from experience the cost of posting a price with a differential greater than the majors are normally willing to allow. The acceptable differential in recent years in market areas where price marketers are relatively strong has ranged from one to three cents per gallon on regular gasoline, with two cents probably being the average. Many price marketers have turned to stamps and premiums as an ambiguous way of extending their price discounts without upsetting markets.

Some of the principal stamp-and-premium-giving price marketers include Martin-Chicago (with 150 stations in sixteen states); Site (with 150 stations in seventeen states); Star (with 163 stations in twenty-two states); and Mars (with 42 stations in eleven states). Customers redeeming stamps at the Martin, Site, Star, and Mars stations during the period of the study could have saved an additional 1.6¢ to 2.9¢ per gallon (see Table 3-4).

TABLE 3-4

VALUE OF STAMPS REDEEMED BY CUSTOMERS

	Stamps to Complete	Normal Rate	Multiples	Purchases[1] to Complete	Cash Value of Book	%	Reduction Cents Per Gallon
Martin	1,000	10/$1	triple	$33.00	$2.00	6.1	2.0
Site	750	5/$1	triple	$25.00	$1.25	5.0	1.6
Mars	130	4/$1	double	$16.25	$1.50	9.2	2.9
Star	90	5/$1	—	$18.00	$1.50	8.3	2.7

1. Assuming an average price of thirty-three cents per gallon.

In addition, two of these stamp givers offered their customers the option of taking premiums in place of stamps. The premiums included such items as glassware, table knives, soap products, and facial tissue.

Merchandise Store Operators

THE merchandise-store price marketers are similar to the stamp-and-premium price marketers. While the stamp givers typically give a cash refund for completed books of stamps, the merchandise-store marketers redeem their stamps for a variety of products.

The mecca for such operations is Minneapolis, the headquarters of Holiday and Super America stations (together operating more than 200 stations). The stations look like large show cases with the merchandise clearly visible from the street (see Plate 3-6). Super America stocks about 1,000 items and offers other items through its catalog. The assortment includes milk, bread, convenience grocery foods, toys, housewares, appliances, film, and auto accessories. Items are priced competitively with nearby supermarkets and discount stores. Super America gives customers large numbers of trading stamps so that they can quickly redeem stamps for products and at the same time offers customers discount prices.[17]

Service-oriented Marketers

INSTEAD of using stamps, premiums, and merchandise as ambiguous means for extending discount prices, some price marketers spend their money to provide unusually good service to their customers. This strategy is unlikely to upset a market. A service-oriented price marketer can sell for two to three cents per gallon less than other price marketers and still give his customers "red carpet" treatment.

While several price marketers strive for good service, few are on a par with the deluxe service given by the Wareco stations headquartered in Jacksonville, Illinois. Wareco uses the "swarm" or "team" technique to service their customers. When a car pulls up to a Wareco pump, a uniformed employee hurries to the island to greet the customer and take the order. A second employee starts dispensing the gasoline and a third cleans the windshield, wipes off the headlights, and checks the oil.[18]

The Wareco service technique combined with its lower prices has resulted in an average of 54,000 gallons per month for the forty-three stations in the chain. This volume is particularly impressive since many stations are located in cities with populations as small as 5,000. Like the other price marketers Wareco has dispersed its operations over several states—Illinois, Iowa, Missouri, and Florida. Claude Ware explains, "We feel [this is] the easiest way to stay out of [a] catastrophe." [19]

In part, Wareco's success is associated with its merchandising skills. Cigarettes and cigars are used as leader items and are sold in amazing quantities. In addition, the Wareco stations cash payroll checks which sometimes run as high as $30,000 between Friday night and Saturday noon. Further diversification of services includes car washes on properties adjacent to fourteen of its stations.[20]

Premium Only

Clark Oil Company is a strikingly successful exception to the trend of marketing multiple grades of gasoline. Prior to February 1971, Clark marketed only one grade—premium gasoline. However, in February, Clark introduced low-lead gasoline in response to the ecology issue. While following its one grade only policy, Clark became one of the largest price marketers. Clark operates 1,500 stations in eleven Mid-continent states.

Clark's success is associated with its well-rounded marketing program. Stations are eye-appealing with the orange, black, and white color scheme carried throughout the operation and are engineered for driver ease and convenience. Uniformed dealers provide quick service. TV advertising has sold the public on the quality of Clark gasoline. Most of the stations are operated twenty-four hours a day.

Clark's policy is generally to price its premium half way between the prices of major-brand regular and ethyl. For example, in St. Louis Clark was selling ethyl gasoline at 37.9¢ per gallon while major-brand regular was priced at 35.9¢ and ethyl was priced at 39.9¢. One of the advantages Clark has with its strategy is that majors tend to price ethyl four cents per gallon over regular while a reasonable cost difference for refining is close to two cents. As a result Clark can discount its gasoline and still have a good margin to use in developing its marketing program. Furthermore, since Clark is not especially price aggressive, it is able to concentrate its operation more than other price marketers and enjoy certain operating economies as a result.

HYBRID PRICE MARKETERS

IN THE YEARS following World War II, several integrated oil companies came to understand that the price-marketer's strategy is based on a sound purchase proposition and they have elected to copy it in several different ways.

Semimajors

THE SEMIMAJORS have characteristics in common with both the price marketers and the major oil companies. They are similar to price marketers in that their dominant concern is selling gasoline at discount prices. As a result the strategy of the semimajors and price marketers parallels one another in terms of location, size, and layout of stations, emphasis on quick service, and price posting.

The semimajors resemble the major oil companies in that both are vertically integrated operations. Most of the semimajors operate terminals, have their own refineries, and produce a portion of their crude-oil needs. However, in terms of size and operating scale they are dwarfed by the major oil companies.

Some of the companies in the semimajor category include American Petrofina (Fina), Kerr McGee (Deep Rock), Apco Oil Company (Apco), Champlain, and Colorado Interstate Gas Company (Derby). Several of these companies represent amalgamations and subsequent mergers of independent crude-oil, refining, and marketing companies. Many of these companies started out as small refiners and crude-oil producers in the South and Southwest. After World War II they were increasingly confronted with problems in selling their refined products. The major oil companies, once their primary customers, purchased less and less gasoline as pipelines were laid to areas formerly supplied by independent refiners. To find markets for their refined products the semimajors increasingly turned to unbranded jobbers. However, in many cases they found that they couldn't profitably compete with the major oil companies for the big-volume private-brand accounts. To survive, the independent refining and crude-oil companies were forced to integrate forward into retailing through the promotion of refinery brands with sales through controlled outlets and branded jobbers.[21]

In one respect, the semimajors differ from both the majors and price marketers. While many of the majors and price marketers distribute their own gasoline through company-controlled service stations, the semimajors rely heavily on independent jobbers, who in turn sell to a large number of independently owned station operators. The semimajors typically offer the independent jobbers and their station owners:

1. lower price than major-brand gasoline;
2. price protection;
3. financial assistance;
4. standardized station image and advertised brand; and
5. company credit-card programs.

Major Oil Companies

Several major integrated oil companies have also begun to enter the price-marketing field. To sell price-discounted gasoline they use secondary brands, many of which they obtained over the past decade and one-half by purchasing price marketers. More recently, the growth of major secondary brands has been from their conversion of marginal branded stations to secondary brands. One major integrated oil company, however, sells its own brand at discounted prices.

Continental Oil Company sells its secondary brands in markets where the Continental brand has little or no representation. For example, in the Southeast and East, Continental owns the Kayo brand (460 stations) and on the West Coast the Douglas brand (476 stations). In addition, on the West Coast a number of majors operate secondary-

and major-brand stations in the same general marketing area. In the vernacular these are known as concubine stations. These include companies like Arco with their Rocket brand, Union with their Hancock brand, Phillips with their Seaside, Blue Goose, and sundry brands. Esso has recently entered the secondary-brand business with its Alert brand. Thus far, the Alert stations are not converted major-brand stations, but rather are entirely new operations.

Hess Oil and Chemical is the only major integrated oil company selling gasoline strictly on a price basis. Hess learned the price marketers' strategy as a supplier and part owner (now 49 percent) of the Meadville Corporation—operator of some of the world's largest volume stations. Hess acquired some of its stations by purchasing those of price marketers and then upgrading the acquired facilities. Hess has also purchased a number of prime locations and built entirely new stations. Where land use is extremely tight, Hess tears down buildings to get the desired locations. Like many other price marketers, Hess specializes in volume gasoline sales to the exclusion of almost everything else—oil changes, TBA sales, lubrication, and repairs. Like the net price marketers Hess does not use trading stamps or games and in some cases has contributed to other price marketers giving up these practices.

Quality of operation is one of Hess's strong points. Hess stations are not only spacious but they are bright, cheerful, and attractive. This effect is created by a yellow, white, and green color scheme. The white uniformed station employees further add to the clean image. Topping off the appearance of station and employees is top quality service—quick, courteous, and complete (see Plate 3-7).

Finally, Hess conspicuously posts its lower price on permanent price signs that are internally lighted and often revolving. As seen in Table 3-5, Hess's prices are considerably below the median major price. In addition, Hess takes the same four cents regular-ethyl differential as the majors.

TABLE 3-5

Washington, D.C. Survey of Hess Oil Company's Prices, Fall 1969

	Regular Gasoline	Ethyl Gasoline
Major-brand median price:	35.9¢	39.9¢
9 Hess stations:	30.9	34.9
	5.0¢	5.0¢

Like most aggressive price marketers, Hess has been able to sell for less by scattering its operation over many markets to reduce its vulnerability to price wars.

Hess's marketing strategy—prime locations, attractive stations, good quality service, and low price—has resulted in exceptionally high-volume stations. From a handful of stations in 1961 Hess has grown in ten years to a firm operating approximately 600 stations averaging 100,000 gallons per month—an average several times the volume of the typical U.S. station.

Rebirth of Self-service Operations

The spectacular history of self-service gasoline merchandising immediately after World War II was short-lived. Urich triggered the boom in self-service in 1947. However, by 1951 the self-service concept had run its course. The final chapter of this early era was a period of prolonged price wars which came to an end with the self-servers giving up their deep cut prices and restoring many customer services (see Appendix B).

Griffin—Pioneer of Self-service

The recent resurgence of self-service operations is in a large part associated with the pioneering efforts of Pat Griffin of Fort Collins, Colorado. A Phillips jobber since 1937, Griffin got his chance to experiment with self-service when Phillips made unbranded gasoline available to their jobbers in 1958. In 1958 he purchased from Vern West of Fort Collins a patented coin mechanism and 16 troublesome coin-operated gasoline pumps. From this point on Griffin quietly and methodically grew. By 1962 he had 136 pumps at thirty stations. These stations sold an average of close to 25,000 gallons per month.[22] By 1969 the number of his stations had grown to ninety-three and they were estimated to be averaging sales of 55,000 gallons per month—more than twice his 1962 average.

Characteristics of Griffin's operation. There are several reasons for Griffin's striking success with his self-service Gasamat operations. One important aspect of the success formula is the patented coin acceptor

Griffin purchased from Vern West and then perfected. This device further reduced the costs of selling self-service gasoline. With coin-operated pumps it was no longer necessary to have an attendant read the gasoline meters and collect money from customers.[23]

Another factor contributing to Gasamat's success is that every Gasamat station is attended. This has proved to be a smart move in several respects.

1. It practically eliminates vandalism and theft.
2. Mechanical breakdowns are quickly noted and repairs made.
3. Attendants can give instruction on the use of coin activated pumps.
4. Safety is increased since personnel are present to oversee the operation.

Furthermore, Griffin has tapped a major unused labor pool to man his stations—senior citizens. Elderly pensioners have all the skills that are needed to supervise the self-service gasoline operation.[24]

Another unique characteristic of Griffin's operation is that the pensioners actually live on the station's premises. Small furnished living quarters are an integral part of each of the Gasamat operations. From within his living quarters, a comfortably seated attendant can observe what is going on and if necessary use control switches to stop the dispensing of gasoline. The "live-in" attendants are paid a commission of under one cent per gallon of gasoline for their services.[25] What Griffin has perfected is a very low-cost method for adequately staffing his self-service stations.

In addition to low labor costs, other aspects of the Gasamat operation are designed to reduce operating costs. Griffin stays away from the high-cost major traffic intersections. He prefers out-of-the-way sites that are inexpensive. Some of his sites are even located back in the residential sections away from the main highway. Griffin has found that given his low prices he doesn't have to locate on "Main Street." His customers will to an extent seek him out. In choosing locations Griffin also tries to stay away from other price competitors to minimize the likelihood of a price war. Furthermore, physical facilities are built for low-cost operation. Many of the original stations had gravel driveways. Inexpensive bright perimeter billboard signs painted with white enamel and lettered in red communicate the savings from trading with Gasamat.[26] As competition in self-service has increased, Griffin has been compelled to up-grade the appearance of his stations. The gravel driveways are now asphalted and the remodeled and new stations are very eye-appealing (see Plate 3-8).

Approximately 100 Gasamat stations are widely scattered through sixty different towns in eleven western states in a manner similar to other aggressive price marketers such as Hudson. While this strategy results in a method of marketing that is relatively expensive to coordinate and supervise, it serves as an effective hedge against disastrous price wars which can destroy a concentrated self-service operation. For example, one price marketer writes:

> Our survival is due in part to a policy formulated years ago: Because of price wars we do not cluster our stations; our policy is one to a town. That is the reason why our operations extend into four states. Those who have grouped their stations in one marketing area are no longer with us. Because of our wide dispersion, all of our stations are never in a price war at the same time.

Gasamat's low operating costs have contributed to the extremely low prices at which its stations sell gasoline. The Gasamat stations generally sell regular grade gasoline for three to six cents less and premium for five to eight cents less than the major oil companies. With these low prices, Griffin has been able to build average station volume to 55,000 gallons per month, while still making a good profit.[27]

Tie-in Self-service Operations

GASAMAT's success has been a significant factor in encouraging others to enter the self-service gasoline business. While many of the new self-service operators have copied features of Griffin's approach, others have used quite different operating methods. One approach has been the self-service tie-in with an existing business.

One of the early companies to promote tie-ins was the Supertron Company, also located in Denver. During 1964 Supertron arranged approximately thirty self-service gasoline tie-ins with convenience grocery stores. In contrast to Griffin's coin-operated pumps, Supertron used a remote-controlled system developed by Herb Timms to regulate the dispensing of gasoline. From inside a store an attendant, utilizing a control mechanism, authorizes customers to purchase specific amounts of gasoline. The customer communicates with the attendant by a two-way speaker system. The gasoline can be sold on either a prepurchase basis (when the customer pays for his groceries) or on a postpurchase basis (after dispensing the gasoline). An adoption of the system allows customers to fill-up their tanks with the amount dispensed shown on the control mechanism in the store.

Tie-in self-service got its biggest boost in 1964 when several large and reputable pump manufacturers began selling remote-controlled coin-operated equipment. From 1964 to early 1969, gasoline tie-ins with food stores have grown from probably less than 100 to an estimated 800 operations. Approximately 500 of these operations are at convenience grocery stores while the remaining 300 are associated with other types of food operations such as dairy drive-ins and hamburger outlets. Hundreds of other outlets such as laundries, liquor stores, and car washes are also tying in with self-service gasoline stations. Among some of the bigger operators promoting self-service gasoline tie-ins in the late sixties were Tenneco Oil with seventy-five convenience store tie-ins; Pioneer Industries, Fort Worth, with sixty-one convenience-store hookups; Farmariss Oil, Hobbs, New Mexico, with fifty tie-ins with ice-cream and root beer stands and convenience stores; and Save More (a Time Oil Company affiliate) of Los Angeles with approximately fifty tie-ins with dairy drive-ins, convenience stores, and other businesses.[28]

Two of the most important factors contributing to the growth of self-service gasoline tie-ins are their modest investment requirements and low operating costs. A self-service gasoline tie-in can be installed for around $10,000. This includes two self-service pumps, underground storage, and remote-control devices. Since tie-in self-service is installed on frequently unused portions of existing retail outlets, the real estate cost is often not computed. Labor costs are treated in much the same manner. Unless a self-service tie-in does unusually high volume, it is anticipated that the existing store help can run the operation so that no labor cost is charged against gasoline. The small investment and negligible labor cost associated with self-service gasoline tie-ins contributes to its also being a very economical method of operation.[29]

Such tie-in operations often sell gasoline from four to five or more cents under the majors' prices and average close to 10,000 gallons per month. This volume compares with a 25,000 to 30,000 gallon average for major-brand stations. While the tie-in volume is not big by comparison, it nonetheless represents plus business to operators. Furthermore, where its growth is not stymied by restrictive legislation, the sheer number of tie-in self-service operations makes it a significant factor in the gasoline business. For example, in Baton Rouge, Louisiana, there were 40 to 45 self-service operations (mostly tie-ins) in early 1969 and it was estimated at least 100 would be in operation within a year. These 100 self-service stations will compete with some 400 major-brand outlets.[30]

Conversions of Traditional Stations

THERE is a third type of self-service operation. This approach does not utilize special equipment. Customers simply dispense their own gasoline and walking attendants collect the money. This is the same method used by the early self-service operators on the West Coast after World War II. The problem with the walking-cashier approach to self-service is that competitors are often unwilling to allow the price differential needed to make the system attractive to customers. Nonetheless, on a scattered basis this approach to self-service seems to be creeping back. As one example, Mary Hudson has converted a majority of her stations to self-service, and most are of the walking-cashier type. By so doing, labor can be cut to one man for each shift. If competitors will allow Hudson and others with operations capable of doing big-volume sales to price four or five cents or more under the major level, then the walking-cashier approach will grow. The reason operations like Gasamat and the tie-ins can have deeply discounted prices is because their costs are so low it is hard to do anything about their prices. It appears that the walking-cashier costs are not quite as low.

Split Operations

SPLIT operations—part full-service and part self-service—are becoming more common. Frequently the split-type station is a defensive undertaking intended to combat straight self-service operations. As self-service gasoline retailing grows more split operations can be expected. On the other hand, the split operation may well be a temporary phenomenon. A major problem with this approach seems to be that the two methods of dispensing gasoline are so fundamentally different that it is difficult for them to coexist in the same station. One of the problems with a split operation is that it is difficult to reduce costs enough to offset the loss of margin from downgrading some customers to the self-service price. Secondly, customer indecision and uncertainty associated with the split operation creates problems for this hybrid operation.

Future for Self-service Retailing

WHILE self-service retailing has been defeated in the past, evidence seems to indicate that there is more permanence to its recent growth. Automated equipment is now playing a central role in self-service operations. As a result operating costs are lower than they were with the earlier self-service experiment. Also, tie-in self-service gasoline operations are diversified so that they are not completely dependent on gasoline revenues in the short run. Both of these conditions make the newer type self-service operations less vulnerable to disastrous price wars than the self-service operations of the late forties and early fifties were. Furthermore, several self-service operators have distributed their stations over a large number of markets, which gives them protection through market and regional diversification.

During the decade of the 1960's self-service gasoline retailing gradually and consistently grew until by 1969 it was estimated to account for approximately 1 percent of the total gasoline sold in the United States. However, from 1969 to 1971, the trend toward self-service operations accelerated and estimates for 1971 have placed sales at 3 to 5 percent of retail gasoline market and stations at 2,500 to 6,800. In those areas where conditions have been favorable for the growth of self-service marketing, it has obtained a sizable share of the market—40 percent in Boise, Idaho; 35 percent in Salt Lake City; 30 percent in Phoenix; and in Arkansas estimates run from 20 to 28 percent.[31]

By mid-1971, thirty-six states approved in some manner the dispensing of self-service gasoline. Do-it-yourself gasoline spread from the near West to the Far West, through the southern part of the country, into the mid-Atlantic area, and on into the New England states. Those states prohibiting self-service gasoline are principally located in the upper Midwest and Great Lakes region.[32] It is anticipated that the ban on self-service gasoline in most of these areas will be broken before long. Barring any fire catastrophes, self-service operations should continue to grow as more areas legalize them, as existing self-servers expand, as new operators enter the field, and as newer concepts of self-service are developed. The strength of the self-service concept is that it is the most economical system in existence for dispensing gasoline and permits gasoline to be sold at the lowest possible price to customers who are willing to do a little work.

Summary

The history of price marketers from the 1930's to the present has consistently demonstrated their resourcefulness in finding ways to reduce costs and lower prices—tracksiding operations, giant self-service stations, volume multipump stations, and restyled self-service stations. Regardless of the particular style of operation, the formula has been one of buying less expensive, unbranded gasoline and selling it through relatively few, highly efficient stations. As a result of the economics of price marketing, gasoline buyers frequently save five cents per gallon of gasoline.

While price marketers do have in common their price orientation, their actual marketing strategies vary greatly. This diversity mirrors the personalities of the founders of these companies and the peculiarities of the markets in which they compete. At one extreme is a group of price marketers who intend to sell gasoline "one cent below everyone else," and who often get their way. Another group of very aggressive but upgraded price marketers runs the limited-service or gasser-type operations that principally pump gasoline and collect money without providing other car service. The net marketer is also a tough price competitor. He doesn't use promotions or gimmicks to cover up his lower prices and often prices under those who do. Stamp-and-premium givers are price marketers who have found it beneficial to reduce the visibility of part of their price cut by using giveaway schemes to pass on savings beyond their posted prices. The service-giving price marketers prefer to pass on additional benefits to their customers by spending money on exceptional services. One successful price marketer sells premium only at two cents under major-brand ethyl.

During the later 1960's and early 1970's new styled self-service gasoline has rapidly grown. Contributing to resurgence of self-service gasoline has been straight self-service type operations, tie-in operations with different types of retailing, and split-pump stations offering both service and self-service gasoline. The economics of do-it-yourself gasoline frequently contributes to self-service gasoline selling for a cent or two less than the service-giving price marketers.

As price marketers have grown in the 1950's and 1960's, integrated companies have become interested in this strategy. These companies have purchased price marketers and built some stations selling a price brand. Those companies entering the price marketing field in this way include (1) semimajors such as Kerr McGee and American Petrofina; (2) major integrated oil companies operating secondary price brands such as Richfield, Union, Phillips, and Continental; and (3) the major integrated oil company selling only gasoline through price outlets.

Price marketers have clearly shown that, if given a chance, gasoline can be sold at low prices on an economy formula, and that the nonprice competitive approach of the major integrated oil companies is not the only way to sell gasoline.

4 / *Competition, Conflict, and Exclusion*

Competition for Markets

Competition for gasoline customers in intermediate to large cities involves hundreds and even thousands of gasoline stations that are all vying for a share of the gasoline market. Depending on the driving pattern of a given motorist, there is a specific subset of stations that are competing with one another for his gasoline purchases. The gasoline station providing the best combination of elements satisfying the needs of a given driver is the one with which he trades.

The competitive offering of different stations is generally based upon variations in products and services sold, free services offered, quality of operation, level of price, type of physical facilities, nature of location, method of payment, advertising support, and sales promotion employed. Much of the difference in the competitive offer of stations is associated with the brand of gasoline sold. Each marketer of gasoline has a general strategy for his brand that he follows in a given market. It is developed by selecting from the large number of costly marketing elements those which a brand management believes will give it the best results for a given level of expenditure (see Figure 4-1). This means that certain elements will be stressed and others will be deemphasized among various brands as they compete with one another.

Intratype and Intertype Competition

Competition among different brands of gasoline can be categorized as either intratype or intertype. Intratype competition occurs among companies with reasonably similar marketing styles—

FIGURE 4-1

Development of a Brand Strategy

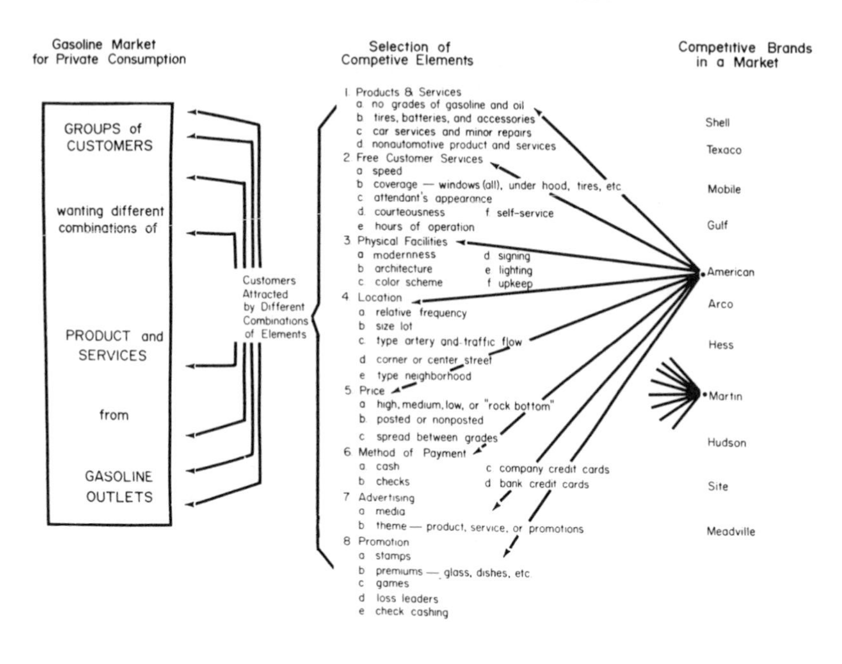

it is a homogeneous type of competition. As discussed in Chapters 2 and 3, intratype competition finds major brand competing with major brand and price marketer competing with price marketer. In contrast, intertype competition is the rivalry between the two principal styles—the price marketers and the majors.

Characteristics of Intratype Competition

While intratype competiton can be strenuous for those involved, it is nonetheless a constrained type of competition. The majors with their top-heavy marketing cost structure and their mutual interdependence are compelled to play down price competition. Similarly, the price marketers, responding to the market void created by the majors, give primary emphasis to lower price and they guard against growing marketing costs that would prevent them from retaining a price edge. Thus, both the majors and price marketers by their own commitments and methods of operating have certain tools to work with and others that cannot easily be used.

There are forces at work in intratype competition which keep those companies pursuing the major marketing style operating pretty much the same way, and those companies following the price marketing approach operating in very similar ways. Because of the basic similarities of intratype competitors, which results from their commitment to a given marketing style, a company cannot allow another intratype competitor to have a competitive edge for long. For example, if one major oil company employs a game and is successful with it, others are going to be forced to engage in the practice, or they can expect to lose some of their position in the market. This is one of the important reasons why all of the major oil companies with the exception of Texaco and Gulf were using one game after another during the game binge of 1967 and 1968. This is in spite of the fact that several major oil company executives were outspokenly critical of the practice. Another example of the intratype pressure to conform is the price uniformity of major brands in markets under normal competitive conditions. Station remodeling is still another example. Let any important major substantially upgrade the appearance of its stations relative to its competitors, and the other majors will soon be compelled to do the same. Similarly, the intratype competition among price marketers forces them to parallel each other's strategic moves. They are particularly sensitive to the price changes of competitors because of the importance of price in their marketing approach.

While intratype competition is constrained, it is nonetheless still rather vigorous. The market fate of intratype competitors is to a great extent determined by the efficiency with which they spend their money on the marketing tools consistent with their market style. During the 1960's there has been considerable shifting in the market position of the majors (see Figure 4-2). Esso fell from first place in gasoline sales in 1960 (9.0 percent share of the market) to third place in 1968 (8.3 percent share of the market). Shell climbed from sixth place in 1960 (6.3 percent share of the market) to second place in 1968 (8.5 percent share of the market).[1] Shell is a good example of a company that has grown by employing its marketing tools in a more effective manner than many of its intratype competitors. Shell led the other majors in closing smaller, marginal stations and replacing them with larger, well-

FIGURE 4-2

ESTIMATES OF MAJOR OIL COMPANY RANK IN GASOLINE SALES AND MARKET SHARES

Company Rank	1960	1961	1962	1963	1964	1965	1966	1967	1968
1	ESSO 9.0	ESSO 8.8	ESSO 8.6	ESSO 8.7	TEXACO 8.4	TEXACO 8.4	TEXACO 8.6	TEXACO 8.6	TEXACO 8.7
2	TEXACO 8.4	TEXACO 8.3	TEXACO 8.1	TEXACO 8.3	ESSO 8.0	ESSO 8.0	ESSO SHELL 8.0	ESSO SHELL 8.3	SHELL 8.5
3	AMERICAN 7.5	AMERICAN 7.3	AMERICAN 7.7	AMERICAN 7.4	AMERICAN 7.7	AMERICAN SHELL 7.7			ESSO 8.3
4	MOBIL 7.0	MOBIL 7.0	GULF 7.0	GULF SHELL 7.0	SHELL 7.5		AMERICAN 7.7	AMERICAN 7.5	AMERICAN GULF 7.5
5	GULF 6.7	GULF SHELL 6.6	MOBIL 6.9		GULF 7.0	GULF 7.2	GULF 7.3	GULF 7.4	
6	SHELL 6.3		SHELL 6.8	MOBIL 6.9	MOBIL 6.9	MOBIL 6.9	MOBIL 6.5	MOBIL 6.6	MOBIL 6.6
7	CHEVRON 4.4	CHEVRON 6.2	CHEVRON 6.0	CHEVRON 6.0	CHEVRON 6.1	CHEVRON 6.1	CHEVRON 6.0	CHEVRON 5.7	CHEVRON 5.7
8	PHILLIPS 4.2	PHILLIPS 4.3	PHILLIPS 4.3	PHILLIPS 4.2	PHILLIPS 4.1	PHILLIPS 4.1	PHILLIPS 4.6	PHILLIPS 5.1	PHILLIPS 5.1

Source: Jeremy Main, "Meanwhile Back at the Gas Pump—A Battle for Markets," Fortune, *June 1969, p. 108.*

located, modern, and attractive stations. Shell also used some of the better marketing games to expose new people to its stations and services, and converted many of them to regular customers. In addition, and not to be discounted in explaining Shell's growth, has been the company's periodic price aggressiveness which has definitely helped it build station volume. Shell's mastering of its marketing tools has contributed to its having the highest per station gasoline volume of the majors.

Similarly, those price marketers who have combined low price with proper emphasis on station appearance, free services, and promotions have gained market position relative to other price marketers. The rundown price marketing stations have lost much of their drawing power. By contrast, "quality" price marketers such as Payless and Golden Imperial have shown substantial growth. For example, the number of stations Payless operated grew from 32 in 1960 to 110 in 1967. This growth record was achieved with new and attractive stations, good service, and low price.

Characteristics of Intertype Competition

Intertype competition differs from intratype competition in that it is concerned with the competitive rivalry between the two principal marketing styles—that of the majors and that of the price marketers—and not simply the vying of similar types of competitors with one another—major opposing major and price marketer against price marketer. This distinction is important since it is intertype competition that shakes the foundations of industries and revitalizes them.[2] In the grocery field, it was the supermarket that challenged and replaced the less efficient and effective neighborhood "Ma-and-Pa" store. Today the supermarket is being challenged by other types of grocery retailing including the discount food store with fewer frills and low prices, and the combination store (supermarkets plus super drugs plus other things) which offers the customer better assortment and saves him costly shopping time. In the general merchandise field—clothing, appliances, housewares—the department store was the undisputed king throughout the 1950's. However, starting with only a very small share of the market in the 1950's, the discount house has grown until it is now number one in general merchandise sales.

Intertype competition is largely responsible for insuring that industries remain strong, efficient, and adaptive to the changing interests of the consuming public. Intertype competition challenges the maintenance of the status quo. One of the factors contributing to the conservatism of the dominant competitors is their financial commitment to a given way of doing business. Another reason why the dominant competitors frequently resist innovation and change is the uncertainty and risk associated with pioneering. Regardless of the reason, marketing innovations typically come from new companies with no vested interest in maintaining the status quo.

There is no reason to expect the major oil companies that have dominated the gasoline business for decades to be any different than the old-line dominant competitors in other fields. The facts indicate that the major oil companies lack progressiveness in their methods of marketing gasoline. When one compares what major-brand gasoline marketing was like in 1950 and in 1970, he is struck by the absence of change relative to what has taken place in other industries. It is true that the average major-brand station has grown larger. In many cases the older stations have been replaced by stations with three or more pump islands and with a larger number of service bays, but the basic institution for selling gasoline and methods used to dispense it have changed little. However, many costly frills have been added including credit cards, trading stamps, and the elaboration of the service station.

THE OTHER SIDE OF THE "RECORD"

EXECUTIVES OF MAJOR OIL COMPANIES and officers of the American Petroleum Institute frequently give talks defending the record of the oil industry. During 1969 and 1970 when many charges were being directed at the industry several of the spokesmen called for presentation of the facts and proclaimed they stood for themselves. The points that were frequently made include the following.

a. From 1948 to 1957 gasoline prices increased only 13 percent versus an increase in the Consumer Price Index (CPI) of 17 percent, and from 1948 to 1969 gasoline prices increased only 23 percent compared to an increase in the CPI of 52 percent.[3]
Untold: The retail price of gasoline (before taxes) increased 51.5 percent from 1946 to 1948.[4] With this huge price increase from 1946 to 1948, oil price comparisons using the 1948 base are heavily biased toward the low side and paint a very favorable picture.

b. From 1957 to April 1969, the consumer price of gasoline increased only about 10 percent, while the CPI increased 28 percent.[5]
Untold: During that same period of time the international price of crude oil fell approximately 40 percent which contributed to a decline in gasoline prices (before taxes) in many European markets. The U.S. was walled off from the lower world crude-oil price to make the U.S. a relatively high-priced market for gasoline.

c. During the decade 1960 through 1969 the petroleum companies averaged 11.7 percent return on their investment versus 12.2 percent for all manufacturing and mining.[6]
Untold: This represents only one set of books for the oil industry which is deflated by extremely liberal tax write-offs and special tax treatment of earnings. Under such circumstances the internal generation of funds through cash flow may be a more important measure of industry performance. In this regard the industry has fared extremely well for until recent years the oil industry internally generated most of

its large capital requirements. Such a feat could not have been accomplished had the members of the industry been doing poorly.

d. *Untold:* The operating margins in the grocery field and general merchandise field for distributing and retailing merchandise actually decreased from 1948 to 1970 as the industries moved from high-cost inefficient methods of selling to mass merchandising. However, in the gasoline industry which clings to the costly service-station approach the cost of distributing and marketing gasoline has continued to increase.

Growth of Price Marketers

Following World War II, intertype competition became very intense. Gasoline purchasers were given more opportunities in many areas to purchase from either the majors or the price marketers. Increasing numbers of drivers elected to purchase private-brand gasoline from the price marketers who operated with fewer frills and who sold gasoline for two to five cents less per gallon than the majors.

While comprehensive figures on the growth of price marketers are not generally available, fragmented information very definitely shows the trend. For example, an executive of Mobil Oil Company stated: "There is no doubt that the emergence of the private brand marketers since World War II has been one of the remarkable developments in the history of the petroleum industry. At least part of their growth has been due to the fact that they have offered to the public lower retail prices than those offered by some major brands." [7]

Detailed information on the growth of price marketers in several states up to 1960 is provided by the Sun Oil Company.

> [The] change in the market position of the so-called private brand retailers of gasoline has been quite marked in the last decade. . . . Surveys made by Sun in 1954 and 1960 showed that private brand stations increased about 67 percent from 4,755 in 1954 to 7,941 in 1960, in those areas where Sun markets. In Connecticut, Florida, and New Jersey the number of private brand outlets more than tripled between 1954 and 1960. In the same period, the number of stations selling the so-called major brands increased by about 2 percent.[8]

More fragmented information on the growth of price marketers comes from several sources.

Minneapolis, Minnesota—By selling gasoline from 1 to 3 cents less than major-brand gasoline, the price marketers were able to increase their market share over an extended period of time.

1935	18.67% share
1940	28.69
1945	26.26
1950	35.31
1951	39.34
1952	42.45[9]

Des Moines, Iowa—Price marketers' share reported to have increased from 9 percent in 1953 to 30 percent in 1956.[10]

Salt Lake City, Utah—The 26 private branders selling for 2¢ per gallon less than the majors were estimated to sell 30 percent of the market in 1963, up from an estimated 18 to 20 percent five years before.[11]

Cobb County, Georgia—A survey by Shell Oil Company indicated that price marketers increased their share of market from 18.05 percent in 1955 to 24.73 in 1958, and to 29.43 percent in 1960.[12]

The timing of the price marketers' expansion in markets varied, depending upon the obstacles that existed.

> In 1957 independents made their first significant market inroads [in Detroit]. The initial success of new entrants, like Martin, Giant, and Clark, encouraged others to enter. With plenty of product available and rapid demand increases, the influx of private branders began to take on invasion proportions. The new private-brand breed combined attractive stations with attractive prices, at least 3¢ below branded posting.[13]

As the report on Detroit indicates the price marketers' penetration of markets occurred at different times and the share that they command in different parts of the country varies greatly. The strongholds for the price marketers have been the Midwest and the Southeast where they held nearly a 30 percent share of the market in the latter 1950's and early 1960's.[14] However, in many other areas where conditions were less favorable to the growth of price marketers their share of the market was frequently insignificant and the public has had no alternative but to purchase major-brand gasoline.

Conflict

The continued growth of the price marketers from the end of World War II through most of the 1950's posed an extremely difficult problem for the majors' style of marketing. If the growth of the price marketers were to continue, and if they were to penetrate new areas of the country, there would be a number of undesirable consequences for the major oil companies. In the short run the majors' share of the market would continue to decline. As this happened the majors would be forced to streamline their marketing and become more efficient until they were in a position to check the growth of the price marketers. However, until that day arrived, the majors would find that they were compelled to sell more unbranded gasoline to continue to run their refineries efficiently and to move their crude oil. Since the unbranded market is very competitive, the return to the major oil companies from unbranded sales would adversely affect their profits.

The threats facing the costly marketing system of the major oil companies had their parallels in other industries. The supermarket and chain-store operators posed a serious threat to the one-time dominant independent grocery wholesaler and retailer. Similarly, the discount store represented a very ominous cloud to the one-time king department store operators. In both industries legal harassment and power techniques were employed in an attempt to check the growth of the lower-cost and lower-price innovators. In the legal field there were local chain-store taxes, blue laws, and retail price maintenance laws to inhibit the growth of the innovators. The principal power technique of the old-line retailers and wholesalers was to threaten or actually boycott manufacturers who sold brand-name merchandise to the innovators. However, in both cases the maneuvers slowed, but did not stop, the inroads of the lower-cost, lower-price, and more efficient retailers.

Similarly, the innovative price marketers represented a severe challenge to the major oil companies' marketing style. Two fundamental courses of action were open to the oil companies. They could either (1) adapt their marketing style to the innovators and become more efficient, or (2) resist the inroads by the lower-cost mass merchandisers by whatever means possible. While both of these approaches have been used, the dominant response of the more important majors has been to fight further inroads of the price marketers. In some cases the techniques used to stop the inroads, and to turn back the price marketers have been subtle. However, in many cases they have been high-handed and obvious.

Exclusion

When companies compete in a free market, the market plays a large role in determining which is going to succeed or to fail. Consumer preference signals and actually determines what type of competitive strategy will gain or lose market position. If a company grows sluggish, then its customers will gradually switch to newer and more progressive competitors from which they derive greater satisfaction.

When competitors receive signals from the market place and they respond, the industry can be characterized as being market directed. On the other hand, if the competitors collectively disregard market signals, the industry can be categorized as being market insensitive. Rather than the market place guiding the major competitors, the collective decision of the industry giants determines how they will compete with one another.

In the following chapters it will be shown that the major oil companies have used a number of exclusion practices to thwart the growth of the innovative price marketers. Some of the techniques used by the major oil companies to "meet the threat" include the following:

1. destructive price wars;
2. price protection and dealer subsidies;
3. "fighting" grades of gasoline;
4. buy-out of price marketers; and
5. increased control over supply.

In most markets where price marketers were important and posed a threat to the major marketing style, all of these techniques were from time to time employed.

The groundwork to head off the price marketers was laid in the latter 1950's and many of the techniques are still employed today. Recognition that the major oil companies were quite upset about the gains of the price marketers came from various sources. For example, the *National Petroleum News Bulletin*, a weekly information service for the oil industry, predicted in 1961 the following: "Things on the selling end are going to get worse. Maybe much worse. The majors are tired of the inroads of the big private branders. To cite a few examples, these non-major marketers enjoy about 25% of the market in Norfolk, 17% in Buffalo, 35% in Indianapolis, as much as 45% in San Antonio." [15] A case in a marketing management textbook that was prepared in cooperation with a large oil company presented the threat of the price marketers in the following way. "During the years following World War II, independent, low-price retailers of gasoline increased their share of the market at the expense of the major oil companies. The independents were so successful that the major oil companies were forced to take note of the independents' method of operation and to *plan specific actions to meet their threat*." [Emphasis added.][16]

5 / *The One-cent Differential Price Wars*

Price Wars—An Abnormal Form of Behavior

To purchasers of gasoline, price wars are viewed as opportunities to buy gasoline at bargain prices for a period of time. An indication to drivers that a price war is building is the appearance of increasing numbers of street price signs at major-brand stations indicating the new and falling price of gasoline. During these periodic price battles gasoline becomes a real bargain, often declining ten cents or more per gallon below the normal price level.

An important question in appraising the consequences of price wars is whether gasoline purchasers benefit from them. The major oil company trade press frequently makes reference to price wars and the low level of prices as an indication of the intensity of competition in the gasoline industry. There is certainly no doubt that gasoline customers in areas with frequent and persistent price wars have benefited from lower prices. For example, in Kansas City, which had been plagued with price wars throughout the 1960's, drivers have purchased gasoline at prices considerably lower than those in other nearby large cities such as St. Louis and Omaha. However, when appraising the consequences of price wars, the short-run considerations are far too simple. What is more important to the public and the economy is the longer-run and broader consequences of price wars for competition. Of particular concern is the impact of price wars on intertype competition.

While gasoline customers are glad to see a price war, this is far from being the attitude of most of the competitors involved. Price wars are actually an abnormal type of behavior[1] involving open industry conflict and not a competitive strategy as described in Chapters 2 and 3. Very quickly prices fall to unprofitable levels of operation and competitors sustain considerable marketing losses.[2] If price wars persist for several months, or are particularly severe, they can financially cripple and destroy the smaller competitors. Under such conditions financial staying power, not operating efficiently, often determines who prevails in these disputes.

Setting the Stage for the One-cent Differential Wars

One of the basic reasons for the dramatic growth of the price marketers following World War II was the price differential between major-brand gasoline and that of the price marketers. In the latter 1940's and throughout most of the 1950's price marketers sold regular-grade gasoline for two to four cents per gallon less than the price of major-brand gasoline.[3] However, as the 1950's progressed, and the price marketers continued to grow, the price differential of two cents or more increasingly became the subject of debate. One of the earlier and more outspoken critics of the two-cents-plus differential was Sun Oil Company's vice president of marketing, Willard W. Wright. Early in the 1950's he questioned the traditional differentials and suggested that there was no justification for a differential of more than one cent per gallon where the price marketers had good acceptance and a good share of the market.[4] By 1958 Wright's position had hardened: "To my mind, . . . the day of the accepted spread . . . is over." [5] Trial balloons were launched to determine the consequences of moving to a one-cent differential. An attack on the traditional differential was made by Pure Oil Company in Birmingham, Alabama, in 1955 with the institution of a one-cent plan and Sun Oil Company followed a year later with a similar plan in Norfolk, Virginia.*

The major oil companies' rationale for narrowing the price differential had little to do with relative efficiency of the majors' and the price marketers' operations. Rather it was based upon the fact that the major oil companies were losing their share of the market as the following statements indicate:

* Both of these plans were challenged by the Federal Trade Commission in 1957 and were dropped several years later at the announcement of the 1965 FTC Hearing on Marketing of Automotive Gasoline.

They're knocking our socks off in too many areas. We can't sit still any longer.[6]

We just have to get in there and mix with them. A 2¢ price differential is ridiculous under today's marketing conditions.[7]

Is the 2¢ differential a thing of the past? Yes, says a number of top executives of major oil companies. This can't continue, . . . differentials must be tightened.[8]

Another point the majors made in arguing for a one-cent differential was the improved quality of the price cutters' operations.

These stations [price marketers] are modern, attractive and in fine locations. The operators advertise . . . the high quality of their products. In fact, they may be better attuned to the times than many long-established marketers and are setting interesting trends in large-volume outlets. . . .[9]

Stanley Learned, president of Phillips Oil Company, dwelled on this point during his lengthy testimony before the FTC's Conference on the Marketing of Automotive Gasoline in 1965.

Thirty years ago most gasoline was marketed through major oil company service stations. The few independent cutrate marketers which existed at that time operated from subpar units, at poor locations, with subquality products and with little, if any, advertising and service. This situation has changed radically over the years. Today these same marketers are operating from representative units in many cases superior to the so-called branded operators, with choice locations, with product quality comparable to other marketers and with similar types of free services. A number of these companies have grown over the years to become major factors in many markets throughout the country. Some of these companies are now operating literally hundreds of stations that are located in many States throughout the Union. This growth has occurred primarily because of the price differential advantage which these people have assumed. . . . [In many cases] the independent refiner and the cutrate marketers are operating from better facilities and better locations than the so-called major dealer, yet, generally speaking, they still expect to have a price advantage of 2 cents a gallon or more simply because they have not been classified as major company dealers.[10]

Shell's One-cent Policy in California

The first broad-scale enforcement of the one-cent differential between the price of regular gasoline at major and price-marketer stations occurred in California. In mid-February 1961 Shell initiated its one-cent plan throughout California which was to ignite many price wars. Shell reduced its prices to within one cent of that of nearby price marketers. The price marketers then reduced their prices to maintain the desired price differentials. On it went until prices had fallen to, and even below, the cost of unbranded gasoline to the price marketers. Under such circumstances operating margins could obviously not cover the costs of doing business. Furthermore, prices were not returned to the vicinity of normal until the price marketers generally accepted the one-cent differential. Enormous losses were recorded by all marketers as the price jockeying went on.[11]

Difference from Earlier One-cent Plans

The Shell one-cent policy supposedly differed from those previously implemented by Pure in Birmingham and Sun Oil in Norfolk. In the Pure and Union test cases the one-cent policy was clearly directed at the price marketers. However, Shell's one-cent policy was professed to be directed at the major secondary brands—in particular at Gulf's recently acquired Wilshire brand. Shell's vice president of marketing, P. C. Thomas, said, "I don't feel that I can give any major a two-cent advantage over me." [12] Shell was quick to point out that it had no policy with regard to the independents. Nonetheless, the effect was the same as Sun's and Pure's one-cent policies aimed directly at the price marketers. Gulf's Wilshire brand was priced at the price marketers' level and Shell was priced one cent above it. The result was that Shell priced its regular gasoline one cent over that of the price marketers.

Shell's one-cent program on the West Coast was watched very closely by other majors. If Shell were successful in enforcing the one-cent differential between the price of major-brand regular gasoline and that of the price marketers, the plan was expected to be adopted in other parts of the country where price marketers were strong. Success with the one-cent policy would mean, for all effective purposes, that the price marketers were no longer going to be much of a threat for without a more significant price difference, the price marketers would lose much of their appeal and many of their customers.

The One-cent Plan in Los Angeles

WHAT Shell's one-cent policy meant and how it was enforced is in part revealed by analyzing the weekly surveys of retail prices of major-brand and price-marketer stations in Los Angeles. Approximately one year after Shell introduced its one-cent plan the bloodiest period of the price war was over and the one-cent policy was reluctantly accepted. A survey of approximately 2,700 stations in the greater Los Angeles area conducted on March 2, 1962, is shown in Table 5-1. Los Angeles is divided into fifteen major communities and the prices are indicated for both the majors and the price marketers. By studying the different areas one will observe that the vast majority of the major-brand stations priced their gasoline one cent above that of the price marketers.

These fifteen communities have been recordered in Table 5-2 to aid the analysis. For the stations located in the first ten communities (1, 4, 10, 2, 7, 9, 6, 5, 3, and 8) more than three-fourths of the major-brand stations in each of the communities were selling regular gasoline at 28.9¢ per gallon and more than three-fourths of the price marketers (with one exception) were selling regular for 27.9¢ per gallon. In the first four zones (1, 4, 10, and 2) involving 485 major-brand stations only one major-brand station was selling regular gasoline below 28.9¢ per gallon and only one of the sixty-six price marketers was selling at below the 27.9¢ price—the one-cent policy was working and holding. In the next six zones (7, 9, 6,5, 3, and 8) the majors' price of 28.9¢ and the price marketers' price of 27.9¢ was holding, but there were a number of off-pattern stations which were creating instability. What the off-pattern figures show is that as some price marketers moved to broaden the spread some of the majors moved to close it.

	28.9¢ or More	−1¢	−2¢	−3¢	−4¢	Total
Major brand:	1,473	53	8	4	1	1,526

	27.9¢ or More	−1¢	−2¢	−3¢	−4¢	Total
Price marketers:	227	45	7	4		283

TABLE 5-1

PRICES OF REGULAR-GRADE GASOLINE BY MAJOR COMMUNITIES IN LOS ANGELES—MARCH 2, 1962

Major Areas in Greater Los Angeles	Majors (M) or Price Marketers (PM)	22.9¢	23.9¢	24.9¢	25.9¢	26.9¢	27.9¢	28.9¢	29.9¢	30.9¢	31.9¢ and More	Totals
1. Valley	M							119	14	2		135
	PM						15					15
2. Burbank and	M						1	113	20	4		138
Glendale	PM						17	3				20
3. Pasadena and	M						18	103	19			140
Alhambra	PM					11	8	1	1			21
4. Santa Monica	M							94	9			103
	PM					1	13					14
5. Hollywood	M						15	290	64		3	372
Metropolitan	PM					11	31	2				44
6. San Gabriel	M						7	119	21	2	1	150
Valley	PM					4	23					27
7. E. Los Angeles	M						1	199	37			237
Bellflower	PM				2	7	61	3				73
8. Inglewood	M			1	3	8	6	310	53	1		382
Huntington Park	PM			3	5	7	58					73
9. Long Beach	M				1		6	214	24			245
	PM			1		5	37	2				45
10. Orange County	M							81	28			109
	PM						17					17

TABLE 5-1 (continued)

Major Areas in Greater Los Angeles	Majors (M) or Price Marketers (PM)	22.9¢	23.9¢	24.9¢	25.9¢	26.9¢	27.9¢	28.9¢	29.9¢	30.9¢	31.9¢ and More	Totals
11. Huntington Beach	M							2	36	2	1	41
	PM						5	5				10
12. Pomona-Ontario	M				64	13	2		1	26	1	107
	PM			18				1	1			20
13. San Bernardino	M			19	25	9	1			21		75
	PM		14	2								16
14. Riverside	M				5					17		22
	PM			2								2
15. Arlington-Corona	M				17	3			1	5	2	28
	PM	1	1	4	1							7
Totals	M			20	115	33	57	1,644	327	80	8	2,284
	PM	1	15	30	8	46	286	17	1			404
% Totals	M			0.9	5.0	1.4	2.5	72.0	14.3	3.5	0.3	100.0%
	PM	0.2	3.7	7.5	2.0	11.4	70.8	4.2	0.2			100.0%

Source: Lundberg Survey Inc., Los Angeles, California.

TABLE 5-2

Prices of Regular-grade Gasoline by Major Communities in Los Angeles, Grouped According to Relative Price Stability—March 2, 1962

Major Areas in Greater Los Angeles	Majors (M) or Price Marketers (PM)	22.9¢	23.9¢	24.9¢	25.9¢	26.9¢	27.9¢	28.9¢	29.9¢	30.9¢	31.9¢ and More	Totals
1. Valley	M							119	14	2		135
	PM						15					15
4. Santa Monica	M	STEADY						94	9			103
	PM					1	13					14
10. Orange County	M							81	28			109
	PM						17					17
2. Burbank and Glendale	M						1	113	20	4		138
	PM						17	3				20
7. E. Los Angeles Bellflower	M						1	199	37			237
	PM				2	7	61	3				73
9. Long Beach	M				1		6	214	24			245
	PM			1		5	37	2				45
6. San Gabriel Valley	M	UNSTABLE					7	119	21	2	1	150
	PM					4	23					27
5. Hollywood Metropolitan	M						15	290	64		3	372
	PM					11	31	2				44
3. Pasadena and Alhambra	M						18	103	19			140
	PM					11	8	1	1			21

TABLE 5-2 (continued)

Major Areas in Greater Los Angeles	Majors (M) or Price Marketers (PM)	22.9¢	23.9¢	24.9¢	25.9¢	26.9¢	27.9¢	28.9¢	29.9¢	30.9¢	31.9¢ and More	Totals
8. Inglewood	M			1	3	8	6	310	53	1		382
Huntington Park	PM			3	5	7	58					73
11. Huntington Beach	M							2	36	2	1	41
	PM						5	5				10
14. Riverside	M				5					17		22
	PM			2								2
15. Arlington-Corona	M				17	3			1	5	2	28
	PM	1	1	4	1							7
12. Pomona-Ontario	M				64	13	2		1	26	1	107
	PM	COLLAPSED		18				1	1			20
13. San Bernardino	M			19	25	9	1			21		75
	PM		14	2								16
Totals	M			20	115	33	57	1,644	327	80	8	2,284
	PM	1	15	30	8	46	286	17	1			404
% Totals	M			0.9	5.0	1.4	2.5	72.0	14.3	3.5	0.3	100.0%
	PM	0.2	3.7	7.5	2.0	11.4	70.8	4.2	0.2			100.0%

Source: Lundberg Survey Inc., Los Angeles, California.

In the last four communities (14, 15, 12, and 13) the markets had collapsed. Instead of the majors selling at 30.9¢ and price marketers at 29.9¢, a majority of the majors were selling at 24.9¢ and 25.9¢ and practically all of the price marketers were selling at 23.9¢ and 24.9¢.

Mechanism for enforcing a one-cent differential policy. As the price marketers reduced their prices in Los Angeles to reestablish differentials of two cents or more, the majors reduced their prices to cut the differential to one cent. As a result of this action and reaction, prices often fell to severely depressed levels where both the price marketers and the major oil companies experienced large financial losses from their marketing operations. After a few weeks of severely depressed prices the major oil companies generally restored prices to the normal level. If the price marketers were willing to maintain a one-cent differential, prices held; otherwise prices started working way down toward the bottom as the price marketers attempted to widen the differential beyond one cent and the majors moved to close it.

Three basic market states were identified in different parts of Los Angeles—steady, unstable, and collapsed. Using the same price survey each of these conditions can be illustrated by one of the communities in the Greater Los Angeles area (see Table 5-3). The steady market is illustrated by the 150 stations in the "Valley." All the 135 major-brand stations were selling gasoline at 28.9¢ or more per gallon and the 15 price marketers were selling it at 27.9¢ per gallon. The market conditions in Pasadena and Alhambra show an unstable area where prices have started to fall. Eleven of the 21 price marketers had reduced their prices to 26.9¢ from 27.9¢ per gallon and 18 of the 140 major-brand stations were pricing gasoline at 27.9¢ rather than 28.9¢ or more per gallon. Shell's role in policing the one-cent differential is apparent; Shell had 13 of the 18 major-brand stations selling gasoline at 27.9¢ per gallon.

The lightning fast speed with which the bottom can fall out of a market is shown by the data on the San Bernardino suburb of Los Angeles. The "last-week" data show that prices were holding constant —all seventy-five of the major-brand stations were at 30.9¢ or more and the sixteen price marketers were at 29.9¢. Someone upset the differential and the market collapsed, falling 6.0¢ per gallon. Within the period of one short week, fourteen of the sixteen price marketers had reduced their prices from 29.9¢ to 23.9¢ and nineteen of the major-

TABLE 5-3

PRICES IN STABLE, FALLING, AND COLLAPSED AREAS IN LOS ANGELES

	Valley					Pasadena and Alhambra					San Bernardino								
Units	27.9¢	28.9¢	29.9¢	30.9¢	Totals	26.9¢	27.9¢	28.9¢	29.9¢	Totals	23.9¢	24.9¢	25.9¢	26.9¢	27.9¢	28.9¢	29.9¢	30.9¢	Totals
Standard		14			14			11		11			7						7
Chevron —		8	8	1	17			9	8	17			6	4	1			9	20
Signal		2			2			3		3			2						2
Shell		28			28		13	12		25		5	1						6
Union		8			8			23	1	24		6	1					2	9
Richfield		12	2	1	15			10	5	15			3					4	7
Mobil		13	3		16		3	12		15		7	2					1	10
Tidewater		7			7		1	13	1	15				5				1	6
Texaco		27	1		28		1	10	4	15		1	3					4	8
Enco																			
Totals		119	14	2	135		18	103	19	140		19	25	9	1			21	75
%		88.15	10.37	1.48	100.00		12.86	73.57	13.57	100.00		25.33	33.34	12.00	1.33			28.00	100.00
Last week		116	17	2	135		16	104	20	140								74	75
%		85.93	12.59	1.48	100.00		11.43	72.28	14.29	100.00								98.67	100.00
Century																			
Douglas						1				1	3								3
Hancock	1				1	1	1			2	2								2
Harbor	2				2		1			1									
Mohawk	3				3														

TABLE 5-3 (continued)

	Valley					Pasadena and Alhambra					San Bernardino								
Units	27.9¢	28.9¢	29.9¢	30.9¢	Totals	26.9¢	27.9¢	28.9¢	29.9¢	Totals	23.9¢	24.9¢	25.9¢	26.9¢	27.9¢	28.9¢	29.9¢	30.9¢	Totals
Powerine						2	1	1		4									
Rocket/Rio							1			1	2								2
Westway	1				1														
Wilshire	3				3	3	1			4	5	1							6
Seaside																			
Veltex	1				1														
Misc.	4				4	4	3		1	8	2	1							3
Totals	15				15	11	8	1	1	21	14	2							16
%	100.00				100.00	52.38	38.10	4.76	4.76	100.00	87.50	12.50							100.00
Last week	15				15	11	8	1	1	21							16		16
%	100.00				100.00	52.38	38.10	4.76	4.76	100.00							100.00		100.00

Source: Retail Gasoline Market Survey, Lundberg Survey Inc., Los Angeles, California, Issue March 2, 1962.

TABLE 5-4

PRICES OF REGULAR-GRADE GASOLINE IN LOS ANGELES 2½ TO 3 YEARS AFTER THE IMPLEMENTATION OF THE ONE-CENT DIFFERENTIAL PLAN

	July 31, 1963								December 29, 1963							
Units	25.9¢	26.9¢	27.9¢	28.9¢	29.9¢	30.9¢	31.9¢	Totals	27.9¢	28.9¢	29.9¢	30.9¢	31.9¢	32.9¢	33.9¢	Totals
Standard						165		165						158		158
Chevron			1		3	364	18	386					4	355	12	371
Signal			1	1	4	71	1	78				1	3	73		77
Shell				2	20	388	1	411				2	15	394		411
Union					1	268		269					1	265		266
Richfield			1			199	7	207					1	197	3	201
Mobil	1			2	8	282	1	294				3	10	285		298
Tidewater			1	4	4	191	2	202			1	2	6	176	2	187
Texaco				2	15	288	2	307				2	12	288		302
Enco						50		50						63		63
American						8		8					3	12		15
Totals	1		4	11	55	2,274	32	2,377			1	10	55	2,266	17	2,349
%	.04		.17	.46	2.31	95.67	1.35	100.00			.04	.43	2.34	96.47	.72	100.00

Units	25.9¢	26.9¢	27.9¢	28.9¢	29.9¢	30.9¢	31.9¢	Totals	27.9¢	28.9¢	29.9¢	30.9¢	31.9¢	32.9¢	33.9¢	Totals
Urich				1	7			8					9			9
Douglas					44			44				4	46			50
Hancock			1	2	54			57				2	46			48

TABLE 5-4 (continued)

Units	July 31, 1963								December 29, 1963							
	25.9¢	26.9¢	27.9¢	28.9¢	29.9¢	30.9¢	31.9¢	Totals	27.9¢	28.9¢	29.9¢	30.9¢	31.9¢	32.9¢	33.9¢	Totals
Harbor					13			13					11			11
Mohawk					17			17					16			16
Powerine		1		2	38			41	1		1	3	38			43
Rocket/Rio		1		1	21			23				2	19			21
Westway					5			5					5			5
Wilshire			2	5	109			116				6	110			116
Seaside					11			11			1	2	10			13
Veltex			2					2			3					3
Misc.	1		13	34	129			177	4	8	18	40	128			198
Totals	1	2	18	45	448			514	5	8	23	59	438			533
%	.19	.39	3.50	8.75	87.16			100.00	.94	1.50	4.31	11.07	82.18			100.00

Source: Lundberg Survey Inc., Los Angeles, California.

brand stations were down to 24.9¢ from 30.9¢ per gallon. The conditions in the San Bernardino area illustrate the process by which Shell and other majors enforced their one-cent differential policy. The price marketers either accepted the one-cent spread or prices quickly fell to an unprofitable level of doing business. After prices fell to very unprofitable levels they normally remained there long enough to discipline the price marketers and then were restored.

Success with the one-cent differential policy. If the price marketers have had enough they maintain the prescribed differential or else the process starts all over again. Apparently, the message—accept the one-cent differential or expect to sell at unprofitable price levels—got across and the price marketers generally, though reluctantly, accepted Shell's plan. Two price surveys in Los Angeles, taken approximately two and one-half and three years after Shell introduced its one-cent program, show a rather remarkable conformity of prices. The survey completed on July 31, 1963, shows that 95.7 percent of the 2,377 major-brand service stations were selling regular-brand gasoline at 30.9¢ per gallon and that 87.2 percent of the 514 price marketers were selling it at 29.9¢ per gallon (see Table 5-4). Five months later prices were moved forward 2.0¢ per gallon with once again rather amazing conformance to the one-cent differential. At that time 96.5 percent of the 2,349 major-brand stations surveyed were selling regular-grade gasoline at 32.9¢ per gallon and 82.2 percent of the 533 price marketers were selling it at 31.9¢ per gallon. The uniformity of prices is truly remarkable when one considers there were nearly 3,000 stations covered by the survey.

Implementation of One-cent Plans in Other Markets

The positive sign coming from the West Coast indicating that Shell was succeeding in enforcing the one-cent plan contributed to its spreading into other areas where price marketers were strong. For example, Texaco initiated its one-cent plan in Wichita, Kansas, where price marketers held an estimated 50 to 60 percent of the market. The pretense used for the one-cent plan on the West Coast was not employed in Wichita for major oil companies did not have secondary brands in this market.

Prior to Christmas 1962 Wichita had "periodic and spasmodic 'price wars' two or three times annually. These 'price wars' were of short duration, seldom longer than 2 or 3 weeks and were caused by various reasons. . . . All of this time . . . the major branded dealers had been retailing their gasoline 2 cents higher than the independents and private branders [the price marketers]." [13]

Texaco introduced its one-cent plan on or about December 26, 1962, by reducing the retail price of gasoline by one cent per gallon and thus narrowing the differential of the majors and price marketers to one cent per gallon. This set off a series of price wars which found the majors narrowing the differential to one cent, the price marketers increasing it to two cents, the majors restoring the one-cent differential, and on it went until prices had fallen seven to ten cents per gallon below normal. There were eighteen such downward spirals from December 26, 1962, through March 15, 1965. In contrast to the West Coast counterparts, the price marketers in Wichita would not hold still for the one-cent plan so market conditions seldom held normal for any period of time and the economic loss to those involved was staggering (more on this later).

The "Wichita story" was repeated throughout the Mid-continent and other areas where price marketers were strong, as majors such as Shell, Texaco, Phillips, and Continental implemented one-cent plans in the early 1960's. For example, Phillips was reported during the severe price wars of 1964 to be enforcing a one-cent policy in 200 markets.[14] The result was a "yo-yoing" of prices as majors made cuts in their prices to reduce the differential to one cent and price marketers cut their prices to restore the differential to two cents. After prices became quite depressed they were periodically restored to see if the price marketers were ready to accept the new differential.

Consequences of the West Coast One-cent Differential Price Wars

Those competitors caught in the pincers' movement of the one-cent differential price wars were severely injured. This includes not only the price marketers (chain marketers, jobbers, and dealers) and the independent refiners selling to them, but also certain regionally concentrated major oil companies and thousands of major-brand dealers and supporting jobbers.

Impact on Major Oil Companies

ON THE SURFACE it might appear that the major oil companies in general would profit from the one-cent differential wars if the price marketers could be corralled. However, as has been indicated it is a costly proposition to enforce the one-cent policy. Those major oil companies concentrated in areas experiencing one-cent price wars suffered along with the price marketers. Tidewater Oil Company was one such company with its marketing properties concentrated on the West Coast and East Coast. In the West Tidewater was exposed to the full force of Shell's one-cent policy and its attendant price wars, which cut deeply into the company's profits. This pricing problem entered importantly into Tidewater's decision to sell its West Coast operation to Humble in the latter part of 1963. When the Justice Department objected Tidewater sought another buyer and ultimately negotiated an agreement to sell its West Coast properties to Phillips Petroleum Company in March 1966.

Two other predominantly West Coast majors sought regional diversity by merging. Union Oil Company announced its plan to merge with Pure Oil Company early in 1965 and later in the year Richfield made known its plan to merge with the Atlantic Refining Company. The newly formed companies were in a better position to withstand devastating regional price wars. Prior to the announced mergers *Forbes* magazine made the following observations about the West Coast.

> Prices were also unfavorable on the West Coast. Standard Oil Co. of California, which controls 25% of the market, joined in fierce price wars in the Los Angeles basin. Thus, despite its big business abroad, Socal's net gained only 4% or so over 1963 on a 5% sales increase. Socal, in effect, has given up more of its marketing profits in order to protect its share of the market.
>
> *West-coast domestics suffered much more* [emphasis added]. Richfield Oil Corp., with only 10% of the market, was expected to report earnings of less than $22 million when the final figures were in (for 1964) off some 20% from 1963. Even though Richfield still ranks No. One among domestics on the Profitability Yardstick, it is last in trend. Tidewater Oil Co.—its attempt to sell its west-coast marketing and refining organization to Jersey Standard thwarted by the Justice Department—struggled through the year with an estimated 7% earnings drop. *Among west-coast marketers, only Union*

Oil Co. of California did well—and not from marketing [emphasis added]. Union's 25% or so earnings increase came from its new production of crude and natural gas in off-shore Louisiana and from bringing in the new Moonie field in Australia.[15]

With regard to the merger of Richfield with Atlantic the *National Petroleum News* made the following observation:

WHAT'S BEHIND THE DEAL

The merger is a very logical one, says marketers close to it. Richfield has heavy representation in California (2,250 of its 3,800 branded stations are there), and is especially strong in the Los Angeles area. But these are highly volatile markets, making Richfield earnings vulnerable to damaging price wars. In addition, Richfield has relatively poor crude-oil sufficiency.[16]

The evidence indicates that one of the consequences of Shell's one-cent policy on the West Coast was increasing industry concentration.

Thousands of major-brand gasoline dealers also suffered as a result of the one-cent differential price wars. As the retail price fell, dealer margins were reduced by as much as 30 to 40 percent. This financial squeeze caused large increases in the dealer turnover rate in the severe price-war areas. During the one-cent price differential wars of the early 1960's large numbers of dealers lost their life's savings. Similarly, major-brand jobbers supplying independent dealers suffered large financial losses.

Impact on Price Marketers

THE ONE-CENT differential plans and the resulting price wars stopped the growth of price marketers and reduced their share of the market in certain parts of the country. The success of Shell's one-cent policy on the West Coast is indicated by the following statements:

After their first full year with a 1¢ differential, many independents are hurting. They don't have the volume they're accustomed to, and they found the price of gimmicking to build volume is high.

* * *

Independents aren't doing well in the wholesale market either. An independent refiner figures they have lost two-thirds of their service-station accounts to majors.[17]

> Until the middle of last year [referring to 1968], independents' share of the market [meaning the five western state area of California, Oregon, Washington, Nevada, and Arizona] had been generally declining in the West since the early Sixties. This period of decline came after the so-called "differential wars," when Shell led an adjustment of the price differential between majors and private brands downward to a normal 1¢ (per) gal.
>
> * * *
>
> Independents wound up in 1968 with a 0.12% increase in market share [to 10.81%] in these five states [—California, Oregon, Washington, Nevada, and Arizona].[18]

In contrast to the 1968 share of 10.81 percent held by the price marketers consider their estimated 1960 share. "The seven majors have roughly 80% of the gasoline market, non-majors (both independent refiners and private-brand marketers) the other 20%.[19] If these two estimates of majors' and the price marketers' share of the market are anywhere near correct, then the price marketers' share has been severely curbed since Shell introduced its one-cent policy early in 1961.

The decreasing representation of price marketers in the Los Angeles market from March 2, 1962, to May 17, 1970, is shown in Table 5-5. The majors operated 85 percent of the stations in Los Angeles in 1962 and the price marketers 15 percent. The price marketers' 15 percent share included the major secondary brands (6.1 percent), West Coast refiner brands (3.3 percent), and the independent marketers (5.6 percent). The table shows very clearly who has grown in the market. The major-brand stations increased by 100 percent over the nine-year period while the price marketers increased only 43 percent. As a result major-brand station representation increased from 85 percent in 1962 to 88.8 percent of all stations in 1970, and the price marketers' station representation fell from 15 percent to 11.2 percent. Unfortunately a station survey could not be obtained for the year just prior to Shell's introduction of the one-cent program. It would very likely have shown an even more precipitous fall in the price marketers' representation. Furthermore, during 1971 while the book was being prepared the Douglas stations were sold and converted to Texaco and the Hancock stations were being disposed of. As a result of these changes two more of the largest price marketers have disappeared from the

TABLE 5-5

STATIONS OPERATED BY MAJORS AND PRICE MARKETERS IN GREATER LOS ANGELES IN 1962 AND 1970

Brand	Number of Stations Operated				Percentage Change from 1962 to 1970		
	March 2, 1962		May 17, 1970				
Majors:							
Socal	587	21.8%	801	15.5%	37%		
Shell	352	13.1	673	13.0	91		
Union-Pure	270	10.0	541	10.5	100		
Texaco	297	11.0	552	10.7	86		
Richfield-Arco	219	8.1	506	9.8	131		
Mobil	285	10.2	566	10.9	106		
Enco (Signal)	69	2.6	288	5.6	308		
American			53	1.0			
Phillips (Tidewater)	205	7.6	302	5.8	47		
Gulf (Wilshire)			302	5.8			
Total:	2,284	(85.0%)	4,584	(88.8%)			100%
Price marketers:							
Owned by Majors—							
Douglas (Continental)	19		53		179%		
Harbor (Union)	21		33		57		
Rocket (Richfield)	26		42		62		
Seaside (Phillips)	16		23		44		
Wilshire (Gulf)	83						
Subtotal:	(165)	6.1%	(151)	2.9%		−8%	
Small West Coast Refiners—							
Hancock	43		87		102%		
Mohawk	12		16		33		
Powerine	33		21		37		
Refiner brands:	(88)	3.3%	(124)	2.4%		41%	
Independents:	151	5.6%	303	5.9		100%	
Total price marketers:	(404)	(15.0%)	(578)	(11.2)			43%
All Stations:	2,686	100.0%	5,162	100.0%			92%

Source: Lundberg Survey Inc., Los Angeles, California, March 2, 1962 and May 17, 1970.

scene further reducing the price marketer's share of the Los Angeles market.

Another unfortuate consequence of Shell's one-cent pricing policy was that it forced many of the price marketers to adopt some type of gimmick to get around the one-cent differential with which they did not feel they could live. These techniques included trading stamps and giveaway discount stamps, which often amounted to another cent or more per gallon of gasoline.[20] In 1960 Blue Chip stamps were introduced on the West Coast, and as Shell enforced the one-cent policy, many of the independents began to give multiple Blue Chip stamps often worth more than one cent per gallon. A public appeal was made by Harry Rothschild, Sr., veteran price marketer-refiner and president of Powerine, to accept the one-cent differential and to make up for this loss by doing the following:

> Use all of the nonprice competitive weapons at your disposal. Give better service; capitalize upon the periods of the day and night when most of your competitors are closed; keep your station area and your personnel looking neater than your competitors; use price leader items such as cigarettes, to keep your traffic flow and sales volume up.[21]

Many of the ideas Rothschild suggested were adopted along with the other techniques mentioned. However, as a consequence of Shell's pricing policy and the action of the other majors, the price marketers were compelled to use less effective marketing tools which contributed to their declining market position.

IMPACT OF MID-CONTINENT ONE-CENT DIFFERENTIAL PRICE WARS ON INDEPENDENTS

THE IMPACT of price wars related to one-cent policies in the Mid-continent also took a heavy toll on the price marketers and independent refineries. The situation in Denver serves as but one account of the implementation of the one-cent plan in the mid-section of the country. The following is a vivid description by the vice president of marketing of Frontier Oil Company of what his company experienced.

> Beginning in 1958 the major price was dropped to within 1 cent of the independent price. Independents, in self-defense to maintain the 2 cents differential essential to their survival, dropped 1 cent in price. Round after round of cuts brought on price war after price war through 1960.
>
> The results were ruinous to independents in the price-war territory—Frontier among them. In 1958 Frontier was serving 87 retail stations in the Denver market. In 1959 it was serving 48, and in 1961 it was serving only 43 stations. By 1963 it had increased the number of stations slightly so that it was serving 50 stations in the Denver area—still less than two-thirds of the number being served when the market was at a 2 cents differential.

But this is only one measure of the losses caused by the 1-cent plan price wars and the resulting low prices. The majors gave price protection to their dealers and jobbers throughout the price-war maneuvers. Frontier with its limited resources was not able to protect its dealers and jobbers in similar fashion for the extended period of the price wars. As a result it and its jobbers suffered. In the year ended October 31, 1960, as compared to the previous year, Frontier lost distribution of over 7 million gallons of gasoline through its jobbers.

At this point, Frontier was forced to surrender. It and another sizable independent refiner accepted a 1-cent differential in an attempt to restore the market to a somewhat near normal basis. The majors have permitted from time to time other independents to post a 2-cent differential.

But the damage was done. Frontier had already lost its jobbers by reason of its inability to continue jobber and dealer margin supports, and the marketing prospects, with Frontier gas selling at only a 1-cent differential, were too dismal for us to recover or replace them. Some of our former jobbers are now jobbers for Continental and Phillips and the rest are jobbers for other majors.[22]

Other independent refiners and marketers were similarly caught in regional price wars that were forcing them into disastrous positions. Thirteen of the primary independent refiners and marketers who banded together to form the Mid-continent Independent Refiners Association petitioned the FTC for relief in 1964. A year later a representative of the group testified that:

. . . This destructive price situation has been most severe in the mid-continent area, where the gross income from refined products during the past two or three years has failed to cover costs of buying and refining crude oil. As a result, the future of independent refining companies, which are unable to finance marketing losses with profits from other operations, is in grave doubts.[23]

Between the time of filing the petition and the hearing a year later some of the refiners were forced to sell out and others shut down refining operations.

The pricing problems in the early 1960's also resulted in the general slowing down and frequent sale of aggressive private-brand marketers. With such chaotic pricing, operations such as Site Oil of St. Louis stopped growing and retrenched in order to survive. Others such as

Star, also headquartered in St. Louis, sold some of their stations to major oil companies to cut their losses. Still others such as Superior and Direct sold their entire operations to the major oil companies. Many smaller price marketers with one or a few stations disappeared completely. They either went branded, sold out, or simply shut down.

Enforcing and Financing One-cent Differential Price Wars

Shell and other majors, initiating the one-cent differential policy, did not simply lower their prices one cent per gallon and allow other competitors to make whatever adjustment they felt necessary. Had the majors reduced their prices in this manner, their actions would have been more consistent with the normal working of competition in which the more efficient reduce their prices to a level dictated by the economics of their operation, and the less efficient either sell at a higher price and gradually lose their business or they lower their prices, but are unable to generate funds to update their facilities and expand, and their share of the market similarly declines. However, the one-cent policy was an entirely different process for it involved deep-cut pricing techniques that had no relationship to operating economies, but primarily involved the use of a financial bludgeon. The technique used to enforce the one-cent policy as previously shown was to drop prices very quickly to levels below the cost of doing business of the price marketers. Often this amounted to price drops of eight cents or more per gallon of gasoline. At such levels marketing losses become staggering for all those operating in the affected area. Unless prices are restored to more reasonable levels those depending entirely, or primarily, on marketing revenues soon find themselves driven to the brink of bankruptcy. However, many of the price marketers reluctantly resigned themselves to the one-cent differential which resulted in a more gradual decline rather than in a short-run catastrophe.

The deep-cut pricing technique is not only used against the financially weaker price marketers, but also effectively applied when major oil companies have disagreements over marketing practices. This is demonstrated by the following exchange which took place among the president of Phillips Petroleum, Stanley Learned, the chairman of the

Federal Trade Commission, Paul Rand Dixon, and a commissioner, Mary Gardner Jones.

Chairman Dixon:	Before we leave this thing about Paraland [an unbranded Phillips' operation selling for 2¢ less than the majors], am I correct in understanding that as of today, you are out of the Paraland operation?
Mr. Learned:	That's right.

* * *

Mrs. Jones:	And are you out because the majors drove you out?
Mr. Learned:	*They drove us out* [emphasis added]. . . .
Mrs. Jones:	How did they do it, by price wars?
Mr. Learned:	By price wars.
Mrs. Jones:	They made it pretty clear to you?
Mr. Learned:	They did not tell us. It was what the market place told us.
Mr. Dixon:	They weren't going to let you sell it for 2 cents less?
Mr. Learned:	We got tired of losing money.
Mrs. Jones:	When you went out they quit?
Mr. Learned:	That's right.[24]

The companies involved in this dispute were Phillips and another oil company with international affiliations. Each of these companies has assets of several billion dollars. In this case Phillips kunckled under; even with its huge size Phillips could not afford to finance the battle.

Phillips was not the only major vertically integrated oil company to highlight the "abnormal" nature of competition associated with the one-cent price wars. George F. Getty II, the president of Tidewater Oil Company, whose West Coast operation had suffered the full impact of Shell's one-cent program made the following statement:

> It is true that many gasoline wars have begun as a result of keen competition. Too often, however, these wars have degenerated into nothing more than competitive cannibalism, with gasoline being sold below cost, and, indeed, in some cases even below the cost of crude oil at the wellhead. The pursuit of this kind of business practice is damaging to the entire oil industry. It forces independent service station dealers and distributors, most of whom are small in size and

> weak in economic power, out of business, and eventually results in the growth of monopoly and in irreparable harm to the consumer.
>
> These "at-any-price" tactics do not improve competition; on the contrary, they destroy it. The situation where a man is driven out of business not because he is inefficient but solely because his larger competitors can afford to absorb below-cost sales for extended periods is, I believe, exactly the type of inequity that Government has a responsibility to prevent.[25]

The second paragraph of Mr. Getty's statement comes very close to being the central point at issue. The "at-any-price tactics do not improve competition; on the contrary, they destroy it." Many companies were driven out of business in the sixties not because they were inefficient, but "solely" because the giants of the industry were able to absorb below-cost sales for extended periods of time.

There is no doubt who were the gainers and losers from the one-cent differential wars on the West Coast. Shell's share of the market in the five western states has been estimated to have increased from 11.7 percent in 1960 to 16.3 percent in 1969. Similarly, Gulf, another of the giant oil companies, lifted its share from a negligible level in the latter 1950's to around 5.0 percent in 1969. While some of Gulf's and Shell's market gains came from other majors (Socal and Tidewater) a very large part of them came from the price marketers. The problem with prolonged price wars, which change the structure of competition as they have on the West Coast, is that the consumer's benefits are short-lived. Once the majors obtain a position of dominance in the market place, the price wars gradually subside and the general level of prices increases.

Some very basic reasons exist why companies such as Shell, Gulf, and Texaco can win their own way when decisions are based upon financial staying power. One of the strengths that the bigger oil companies have is their regional diversity. Since generally they have good representation in the Northeast, the mid-Atlantic, the Southeast, the Great Lakes region, the South and the Southwest, and the West they can withstand for extended periods of time depressed price conditions in certain parts of the country. They have the important element of time in their favor as they seek to extend their position in certain markets. The regional fluctuation of earnings is indicated in reports such as the following.

> With all the principal reports in [the NPN] BULLETIN's tabulation of 27 companies shows 15 up over first-half 1961, with 12 down. Biggest gainers were companies based on the *West Coast*, where gasoline prices have stabilized after turmoil a year ago. Weakest showings were in the *Midwest* and *Southwest*, where subregular gasolines have shaken up markets. Four of the five "*international*" majors registered solid gains.
>
> *Jersey Standard* and *Texaco* stood one-two in size of net earnings, as usual. *Standard of California* moved up a notch into third place, dropping *Gulf* to fourth. *Mobil* continued in fifth. *Shell* regained sixth from Indiana Standard, after being edged out a year ago. [Emphasis added.][26]

The report indicates another factor that gives certain oil companies great financial staying power. They are giant international oil companies. As the report states, the five oil companies in 1962 with the greatest earnings were the internationals—Standard Oil of New Jersey, Texaco, Standard Oil of California, Gulf, and Mobil. In sixth place was Shell which is an independently run subsidiary of the international Royal Dutch Shell Company. These companies, in addition to area diversification, have international diversification which gives them even greater ability to withstand periodically depressed prices. For example, during the first half of the 1960's the international oil companies had outstanding earnings abroad which helped them to offset poorer domestic earnings. This is shown in Figure 5-1.

Another factor that gives certain large oil companies tremendous financial power is that they are vertically integrated. Marketing is like the exposed part of the iceberg and only involves approximately 20 percent of oil companies' total investment. However, for many of the price marketers a greater proportion, if not all, of their assets are involved in marketing. Should a major oil company with a strong crude-oil position operate its marketing facilities in certain areas with little or no return on investment, it would have very little, if any perceptible impact on overall profits. This would be particularly true if a major oil company's crude-oil throughput increased significantly in the depressed price areas. However, for those price marketers with a much higher percentage, and possibly all of their assets in marketing facilities, it would have a major and lasting effect were they to operate with little or no return on their marketing investments for a prolonged

FIGURE 5-1

Return on Average Invested Capital

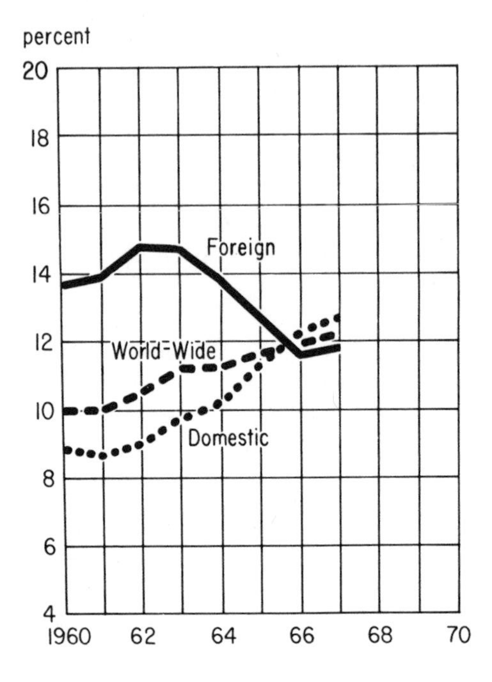

Source: "*Annual Financial Analysis of a Group of Petroleum Companies 1967*" (*New York City, Chase Manhattan Bank, 1968*), *p. 21.*

period of time. Furthermore, several of the major oil companies have diversified into other fields such as petro-chemicals and different types of energy such as coal and uranium which gives them still more income stability.

6 / *Subregular-grade Gasoline Price Wars*

SHORTLY AFTER the beginning of the one-cent differential price wars, the introduction of lower-octane, subregular-grade gasoline triggered in addition "tane" price wars throughout large sections of the country. The tane price wars were very similar to the one-cent differential wars in terms of those competitors principally involved, the objectives of the war, the techniques employed, and the outcomes.

BACKGROUND OF NEW GRADES OF GASOLINE

FOR THE FIRST HALF OF THE 1950's practically all of the major oil companies were selling two grades of gasoline—regular and premium. Regular grade was designed for older cars with relatively low compression engines and smaller intermediate model new cars, while ethyl was developed for new and bigger model automobiles with higher performance engines.

The exception to the two-grade practice was Sun Oil Company which sold only one grade of gasoline—Blue Sunoco. However, Sun found that its competitive position with one grade of gasoline was weakening as Detroit moved increasingly toward higher performance engines with greater compression ratios. If Sun were to continue to produce one grade of gasoline good enough for most cars, production costs would price it out of the lower-octane, and less expensive, gasoline market. On the other hand, if Sun were to produce one economical grade of gasoline for cars having middle-range to lower-octane requirements, the product would not satisfy the needs of the newer higher compression cars. Sun concluded that to protect its position a new grade strategy had to be developed. While Sun could have moved quickly to a two-grade operation, management wanted, if possible, to develop a new approach to take advantage of the increasing range of automobile octane requirements. Early in 1956 Sun conducted a test of a blend pump in seventeen Orlando, Florida, stations. This was essentially a "dial-a-grade" approach which gave drivers the option of buying five grades of gasoline from a single pump with Blue Sunoco being sold from a separate pump. The six grades were designated as Sunoco 200, 210, and on to 250. Sunoco 200 was equivalent to regular gasoline and was rated as 93 octane research method while Sunoco 250 was a super premium gasoline (having a higher octane rating than most "ethyl" grade gasolines). The other grades had stepped up octane ratings that came between Sunoco 200 and 250.[1] Within a year of Sun's initial test, the blend pump was introduced throughout Florida. Sun then announced its plans to spend $30 million to extend the "dial-a-grade" system throughout its twenty state marketing area during 1957.

The first company, however, to introduce a "third-grade" gasoline throughout its operation was not Sun Oil, but rather Esso. To meet the projected increases in automobile octane requirements Esso decided to introduce a superpremium-grade gasoline—Golden Esso Extra. In July 1956 Esso started converting its operation to three grades of gasoline and by mid-August it was estimated that 75 percent of Esso stations were selling Golden Esso Extra. During the first six months Golden Esso Extra was estimated to account for approximately 10 percent of dealer sales. Gulf Oil Company became the second major oil company to commit itself to sell a third-grade superpremium gasoline. Gulf Crest was introduced at 10,000 stations April 15, 1957, and was priced three cents per gallon above ethyl.

By 1958 most of the major oil companies had taken a position with regard to three or more grades of gasoline. Of the twenty largest, only Standard Oil of New Jersey, Standard Oil of California, Gulf, and Sun Oil decided to market three or more grades of gasoline. The other oil companies such as Texaco, Mobil, Shell, and Phillips decided not to make the investment in three grades, but rather to stay with two grades. To satisfy the octane requirement of the bigger cars, the two graders planned to increase the octane levels of their premium gasoline.[2]

Introduction of Subregular and Middle Grades of Gasolines

A new twist was added to the multiple-grade gasoline development when Sun Oil tested a "cent-a-grade" station in New Castle, Delaware, late in 1960. There were nine grades of gasoline priced one cent apart. Two of the grades, Sunoco 190 and 180, were subregular grades and priced one cent and two cents below the company's regular, Sunoco 200. Not unexpectedly the station was tested in a "hotly competitive area with many independents" for Sun had been searching for a strategy to follow with regard to the price marketers.[3] Sun was apparently satisfied with the results from the one-station test and a broader scale test was designed. On May 1, 1961, the "nine-grade" program was introduced at the fifteen Sunoco stations in Charlotte, North Carolina. The lowest grade, Sun 180, was priced "to meet the independents on the nose." [4]

Gulf Oil Company started a test program similar to that of Sun. Gulftane, a low-octane, subregular-grade gasoline, was inaugurated in San Antonio, Texas, in July. On August 4 Gulf extended the test to Indianapolis, Indiana, and from there to Buffalo, New York, and Norfolk, Virginia. In three of the four markets Gulftane replaced the slow moving superpremium grade Gulf Crest, while in San Antonio where Gulf Crest had not been introduced equipment had to be added to accommodate a third grade of gasoline. Esso, the third major oil company with multiple grades, also found sales of superpremium slow and decided to test a new strategy. In contrast to Gulf's strategy of a subregular grade, Esso decided to introduce a "middle"-grade gasoline. This resulted in three grades, Esso regular, Esso Plus (middle grade), and Esso Extra (premium).

The reason for Esso's and Gulf's experimenting with new third grades of gasoline to replace their superpremium grades was that sales volumes had not lived up to expectations. Instead of the prediction that Detroit would build bigger and more powerful cars requiring higher-octane gasoline, the car manufacturers changed their strategy. Detroit began to produce the economy "compact cars," which did not even require the octane ratings of regular gasoline.[5]

Primary Target of Subregular-grade Gasoline

Gulf's and Sun's introduction of subregular grades of gasoline was to a large extent prompted by the growth of the price marketers in the previous decade. As previously discussed, Sun's vice president of marketing wanted to reduce the price differential of the majors and price marketers from two cents to one cent on regular gasoline because of the growth of independents in Sun's marketing territory. With regard to Gulf's reason for introducing Gulftane, *Forbes* magazine said that the new grade of gasoline drew "a bead on the independents, . . . the fastest-growing chunk of the gasoline market: the estimated 2 million barrels that are sold daily in the U.S. for *2¢ less* a gallon than the major companies' regular brands." [6] The same article quoted Gulf's chairman, William K. Whiteford, as saying, "Private brands can murder the majors with a 2-cent or 3-cent spread. They once had 5% of the business; now they have 25%. We *can't continue to let them sell the same quality of gasoline for 2 cents or 3 cents less*" (emphasis added). Furthermore, it was no coincidence that the four markets Gulf chose for testing Gulftane were "strongholds" of the price marketers. Estimates of the price marketers' share in these markets are indicated below.[7]

San Antonio	45%
Indianapolis	35%
Buffalo	17%
Norfolk	25%

Market Reaction to Tests of Subregular Gasoline

In the five test markets—Charlotte, San Antonio, Indianapolis, Buffalo, and Norfolk—the immediate consequence of introducing the subregular grades of gasoline was the same. Prices quickly dropped to below cost levels with those involved experiencing huge marketing losses. What the tanes did to prices in the test markets in which they were introduced is shown in Table 6-1. In four of the five test markets (Buffalo, Indianapolis, Norfolk, and San Antonio),

TABLE 6-1

EFFECT OF INTRODUCTION OF SUBREGULAR-GRADE GASOLINE ON MARKET PRICES

Market	Majors (Regular)	Gulftane	Sun 180	Private Brands	Tax
↓ Buffalo	11.6¢–9.6¢	9.6¢–8.6¢	9.6¢–8.6¢	9.6¢–7.6¢	10¢
↑ Charlotte	20.9¢	—	18.9¢	18.9¢	11¢
↓ Chicago	23.9¢–20.9¢	—	—	20.9¢–17.9¢	9¢
↑ Dallas	20.9¢	—	—	18.9¢	9¢
↓ Detroit	18.9¢	—	—	16.9¢	10¢
↓ Indianapolis	13.9¢–12.9¢	12.9¢	12.9¢	12.9¢	10¢
↓ Ft. Worth	16.9¢–15.9¢	—	—	14.9¢–13.9¢	9¢
↓ Norfolk-Newport News	10.9¢–9.9¢	9.9¢–8.9¢	8.9¢	8.9¢–7.9¢	11¢
↓ Oklahoma City	20.4¢	—	—	18.4¢	10½¢
↓ Providence	15.9¢–14.9¢	—	—	13.9¢	11¢
↓ San Antonio	10.9¢	9.9¢	—	9.9¢	9¢
↑ St. Louis	22.4¢	—	—	20.4¢	8½¢
↓ Tulsa	20.4¢	—	—	18.4¢	10½¢
↓ Waco	8.9¢	—	—	6.9¢	9¢

Source: National Petroleum News Bulletin (*September 11, 1961*), *p. 2.*

major-brand regular gasoline fell to approximately 10¢ per gallon below what had been considered normal in these markets.

The process by which prices fell to the low levels shown was often as follows:

1. Gulftane would match the low independent price.
2. Other majors such as Mobil would match Gulftane prices.
3. The price marketers would lower their prices one or two cents below Gulftane to restore the differential.
4. Through action and reaction prices would fall to the bottom, eight to ten cents below normal, and hold there for a period of time as huge marketing losses were accumulated. Gulftane was priced with the independents' regular and most majors' regular was a cent above.
5. Eventually prices would be restored. If the price marketers would allow Gulftane to sell even with their regular-brand price, and the majors would sell their regular for a penny more, the markets might hold normal for a period of time.
6. Within a few days or weeks following restoration markets would typically start to collapse as different companies deviated from the new conditions prescribed for price stability.

Price Gyration in Indianapolis

AN EXAMPLE of this process can be observed from what happened when Gulftane was introduced in Indianapolis, Indiana, on August 4, 1961. The account of what took place shows how the price of major-brand regular gasoline was reduced from 31.9¢ to 22.9¢ per gallon in ten days, followed by a partial price restoration before prices fell back to 22.9¢.

> Friday morning, August 4, 1961, all Gulf stations in Marion County were posting a large price sign of 28.9. Previous to this time majors were selling 31.9 for regular and 35.9 Ethyl, Independents 28.9 regular and 30.9 Ethyl.
>
> Hudson moved to 26.9 Friday afternoon and was followed by Gulf. Saturday afternoon Aloha moved to 24.9 followed by Gulf, Hudson, Owens, Payless, etc. Monday majors moved to 26.9 and by afternoon Hudson, Aloha, etc. were 22.9 followed by Gulf at 22.9. Majors moved to 24.9.
>
> Wednesday Mobil moved to 22.9 followed by Shell, Phillips and on Friday afternoon Standard moved to 22.9.
>
> On Monday, August 14th all independent and majors at 22.9; two Gulfs at 21.9; one Shell and one Texaco and a Marathon at 21.9.
>
> The weekend of August 19 to 20, found both major oil companies and private brands selling regular at 22.9. There were three stations selling Gulftane at 21.9 and several private brands at a 20.9. One private brand held a 25.9 until Friday, then took a 18.9 over the weekend.
>
> Monday morning, August 21st at 8 A.M., Gulf began moving Gulftane up 3¢ to 25.9. In the afternoon several private brands, Wake Up, United, Imperial, Petroleum System, moved to 25.9. Texaco moved to 26.9.
>
> Tuesday, August 22nd, most majors moved to 25.9, Shell to 26.9. Most private brands moved to 25.9, including Hudson.
>
> Wednesday, August 23rd, the few remaining majors moved to 25.9. One private brand on 25.9 moved to 22.9, then up to 23.9. Site moved up 1¢ to 23.9, Kocolene moved to 23.9.
>
> On Thursday, August 24th, most private brands started moving to 23.9, 2¢ under the major posted price.

Friday, August 25th, found private brands moving or already on 23.9. In the afternoon Gulf moved Gulftane down 2¢ to 23.9.

There were exceptions on the posted price by major stations but this was caused by the individual station lessee.

Several outlying towns up to 40 miles away have private brands at 22.9 or 24.9, majors at 24.9 to 26.9.

Saturday, August 26th, Hudson and Site to 22.9.

Monday, August 28th, Gulftane to 22.9, some stayed on 23.9.

Friday, September 1st, Mobil moved to 22.9, dealers did not make move till 9/5/61, day after Labor Day.

September 11th, private branders and majors at 22.9 or 23.9, including Gulftane.[8]

Conflicting Positions Over Pricing Subregular-grade Gasoline

The reaction of both majors and price marketers to the introduction of Gulftane was not surprising. The majors' position seemed to be basically that they could not permit Gulftane to sell in excess of one cent under their regular-grade gasolines, since Gulf was a recognized brand with a strong credit-card program and Gulftane satisfied the requirements of most regular-grade gasoline customers. The price marketers' argument paralleled that of the majors, but from a different viewpoint. Their position was that Gulftane must sell for one cent more than their regular-grade gasolines, since Gulf was a recognized brand with a strong credit-card program and Gulftane satisfied the requirements of most regular-grade gasoline customers. Finally, Gulf's position was that Gulftane would be priced the same as the price marketers' regular-grade gasoline.

The conditions that the combined actions of the majors were prescribing for market peace and normal prices were similar to those terms associated with the West Coast price wars which were taking place at the same time. The price marketers would have to accept a one-cent differential between the price of their regular gasoline and that of the majors. However, there was one new twist and that was that Gulftane would be sold at the same price as the independents' regular gasoline. The one-cent differential plus the introduction of Gulftane were basically unacceptable to the price marketers. This impasse set the conditions for the massive price wars that were to follow.

There were other reasons that companies reacted so strongly to Gulftane. First, various estimates of those people familiar with oil technology put the cost of producing Gulftane (supposedly a 91-octane gasoline) approximately one-half cent per gallon less than Gulf regular (approximately a 93-octane gasoline).[9] Nonetheless Gulf was frequently selling Gulftane one to two cents less than its regular. In essence Gulf was taking a marginally lower-quality product and selling it for considerably less than its pricing policy on regular. Thus, Gulftane was being used as a "fighting" or a "combat" grade of gasoline. Second, there was a discrepancy in what Gulf told competitors and what they told the public about Gulftane. To competitors Gulf implied that it was merely producing a lower-quality product which would more economically satisfy the requirements of the compact economy cars.[10] However, Gulf did not promote its product as being designed for compacts, but on the other hand advertised Gulftane as satisfying the octane requirement of 50 percent of the cars on the road.[11] This put Gulftane squarely in competition with the price marketers and majors for regular-grade gasoline customers.

Another factor that made the price marketers suspicious of Gulf's motives was that when Gulftane was introduced into Indianapolis it was "consigned" by Gulf to its dealers "so Gulf could change prices [of its independent dealers] at will without fear of prosecution by their dealers for unfair pricing practices." [12] When Gulf's field representatives found lower prices charged by other dealers they would lower Gulftane's price to meet them immediately. Thus the price marketers came to quickly realize that they were not competing with independent dealers, but rather with the huge Gulf Oil Company.

Decision to Market Subregular-grade Gasoline

Gulf announced that after a ninety-day test of Gulftane in San Antonio, Indianapolis, Buffalo, and Norfolk a computer would decide the fate of Gulftane. "We don't care what anybody thinks—we don't want to know what anybody thinks. The machine is going to tell us if we are right or wrong."[13] Apparently, the computer signaled the go ahead for Gulf's chairman, William K. Whitehead, ordered Gulftane into most of its 36,000 stations in thirty-nine states in November 1961.

Similarly, Sun Oil announced in January 1962 that it would sell one subregular grade, Blue Sunoco 190, through its 8,700 stations in twenty-four states. In February 1962, Standard of Kentucky announced as a defensive move, that it would introduce a subregular grade in its five-state southeastern market. These were the only major oil companies to adopt a subregular grade of gasoline during the first half of the 1960's.

One cannot help but wonder about the computer analysis that Gulf made of the test markets. From a volume standpoint there would be no doubt about an improvement in Gulf's position. However, there must have been no profit analysis, or else it was ignored, for during a good part of the test period, if not a majority of the time, gasoline was being sold at, or near, refinery prices for branded gasoline. Marketing costs either were not being covered at all or were being covered only in a small way. As a result, Gulf's decision to introduce Gulftane throughout its more than 30,000 stations must have been predicated on the belief that there would be safety in numbers and that competitors could not afford to fight Gulftane on a broad-scale basis in contrast to what had happened in the four test markets.

Why did Gulf, the third largest oil company in number of stations, with international operations, become involved in such a potentially market-disturbing proposition as Gulftane? A partial answer seems to be that Gulf had some serious marketing problems that required a dramatic and unorthodox approach such as a subregular "fighting" grade of gasoline. The specifics of the situation Gulf was confronted with included the following:

1. Gulf had engaged in a massive two-year station building program involving 4,200 stations that had commenced in 1956. Many of the stations were closed for long periods of time and for others satisfactory throughput had not been developed.
2. Gulf like others had been losing gasoline volume. Four years before Gulf lost a large supply contract with Spur, then an aggressive price marketer. During the first three-quarters of 1961 volume was reported off 6 percent from the year before.
3. The price marketers had been growing and were cutting into the major oil companies' business.

4. Gulf Crest was not reaching its objectives and Gulf had an opportunity to develop a different approach.[14]

It also appeared that Gulf's problems were related to marketing operations that were often below par with respect to other majors. For example, frequently its locations were not particularly good, its station design and color scheme lacked appeal, and often its dealers were not of the quality of those of other majors. As a result, a lower-priced product seemed to be a short-run approach by which Gulf might work its way out of its problems.

Spread of Tane Price Wars

As Gulftane rolled into more markets in November and December of 1961 and in the early part of 1962, the result was often severe price wars as majors and price marketers fought over differentials. According to the earnings' reports, hardest hit by the price wars resulting from the subregular grades of gasoline were dealers in the Midwest and Southwest.[15]

For example, consider the following report.

> "WORST EVER," "NO HOPE IN SIGHT,"
> SAYS EMBATTLED MIDWESTERN MARKETERS
>
> The Midwest is a "critical one" says a veteran marketing executive. St. Louis and Detroit stand out as sorely troubled spots. Also bad is Ft. Worth-Dallas Market.[16]

The *Bulletin* went on to state that major-brand prices in St. Louis, Detroit, and Denver were off by eight to ten cents per gallon. As noted earlier these were the areas where the price marketers either had significant market positions or were making inroads. In contrast, in parts of the East where the price marketers' share was relatively small Gulftane and Sun 190 were reported not to be upsetting markets nearly to the extent that they were in the Midwest.

There is little doubt about the devastating effect that Gulftane had on prices in areas where price marketers were concentrated. Gulftane insisted on pricing with the independents' regular and the other majors were determined to be one cent above. The price marketers were expected to accept a one-cent differential between their regular price and that of the majors if there was to be any market peace. If not, prices would be taken down far below the independents' cost of doing business and held there for a prolonged period of time. In this sense the tane wars were similar to the one-cent differential wars of the West Coast. Deep-cut, below-cost prices were being used to enforce the one-cent differential with Gulftane at the independent level. Thus, the massive financial strength of the major oil companies was being used to regulate the price marketers' pricing policy.

Attempted Price Restorations

APPROXIMATELY ONE year after Gulftane was introduced, Gulf started a series of national price restorations, hoping that price marketers and majors were ready to have peace. A news release by Gulf Oil pointed out the price disturbances associated with Gulftane. "On August 27, due to severe and prolonged price disturbances throughout Gulf's marketing territory, our company removed all dealer assistance on gasoline with the hope that the market would permit our dealers to enjoy normal margins of profit." [17] For three weeks following the restoration Gulf suspended the practice of meeting the price marketers' regular with its Gulftane, and held Gulftane one cent below its regular, regardless of the price marketers' postings. In some cases the price marketers advanced their regular price to the Gulftane levels, others took a position of one cent under Gulftane, and a hard core of aggressive marketers held to two cents under Gulftane.[18]

Starting on September 15, 1962, three weeks following Gulf's restoration of prices to normal, Gulf announced that it would begin to grant dealer allowances on Gulftane so that dealers could become competitive again. As Gulftane prices were again lowered to match those of the price marketers, the question became, what would be the response of the two-grade major oil companies? If they stuck with their position of pricing one cent over Gulftane, prices would quickly work their way back to the bottom in many markets.

The response was varied. In some markets the price differential between the majors' and the price marketers' regular gasoline expanded to three and four cents with Gulftane at the price marketers' level. Such conditions tended to erode the restoration with increasing numbers of markets falling off normal. In some areas where the price marketers accepted the one-cent differential, they restored some of the price edge lost on regular by reducing the price of premium gasoline to one cent above their regular, making their premium price equal to major-brand regular. For example, the price marketers' regular might be 29.9¢ and ethyl 30.9¢ with the majors' regular 30.9¢ and premium 34.9¢ per gallon. This typically meant that the price marketers had a one-cent differential on regular and a four-cent differential on premium.[19]

The deterioration of markets following the August 29th nationwide price restorations led Gulf two months later, in early November, to remove price support once again and to restore prices. By mid-December Gulf was restoring support as price conditions in markets started deteriorating all over again. For the third time Gulf withdrew support and nationally restored prices in mid-April 1963. Whether the restoration would hold depended upon much the same questions as in the past. Would the price marketers accept the one-cent differential or were the majors prepared to permit a two-cent differential? Within three weeks the same unsettling conditions were appearing once again. Since many price marketers were not coming up to the level of Gulftane's prices, Gulf started lowering Gulftane's prices and markets began to erode.[20]

This condition of rapidly deteriorating prices in certain markets is shown by the prices in Table 6-2, which gives the general prices in

TABLE 6-2

RAPID EROSION OF PRICES FOLLOWING RESTORATION IN SELECT MARKETS

		Major Regular	Private-brand Regular
Wichita:	4/4	16.9¢ (minus tax)	14.9–15.9¢ (minus tax)
	4/19	20.9	18.9
	5/3	14.9	13.9
Detroit:	4/4	14.9	12.9
	4/24	22.9	19.9–20.9
	5/7	17.9	15.9
St. Louis:	4/8	13.9	11.9
	4/25	21.9	17.9–18.9
	5/3	17.9	14.9
	5/5	16.9	14.9
Miami:	4/2	13.9	12.9
	4/25	20.9	19.9
	5/8	14.9	13.9

Source: "Gasoline Prices Weaken More, Approach Pre-Price-Hike Levels," National Petroleum News Bulletin (*May 13, 1963*), *p. 1.*

four markets two weeks before the restoration, during the week of the restoration, and two weeks after it.

Modification of Gulf's pricing policy. In August 1963, approximately two years after the introduction of Gulftane, Gulf announced that it would accept a one-cent spread between the price marketers' regular and Gulftane in some markets. During the preceding two years Gulf's policy had been to meet the price marketers' regular-grade gasoline price with Gulftane. Now Gulf was taking the position that it would split the difference—one cent over the price marketers and one cent under major-brand regular. A Gulf spokesman said the objective was to help stabilize some markets.[21] The question then remained, what would other majors do? Would they be willing to accept a two-cent differential? It soon became apparent that majors in some markets were unwilling to spot the price marketers more than one cent and to price "on the nose during a price war." [22]

Manipulations of Price Protection

In some markets the price aggressiveness of certain majors created a "Frankenstein" monster. Now that the majors generally wanted to get prices back under control certain major-brand dealers were not cooperating. Instead they were playing the market for its ups and downs—buying large inventories when gasoline was cheap and then undercutting other majors when prices were restored. This kept some markets in a continuous price "yo-yo." One such market was Detroit which not only had price marketers and a large number of stations selling tanes, but also many major-brand "maverick dealers." The dealer's price of gasoline fell to 6.8¢ per gallon during the summer of 1964. At this low level the dealer's price did not even cover the posted price of the crude oil from which the gasoline was made and made no contribution to refining and marketing costs.[23]

Severe Depression of Wholesale Gasoline Market

The latter part of 1964 did not see pricing problems easing in the Southeast and Mid-continent as some earlier signs had indicated they might. The unbranded price of gasoline dropped to an eighteen-year low at the Gulf and fifteen-year low in the Mid-continent. Independent refiners' prices were so low as to all but eliminate, or actually eliminate, refiners' margins.[24] The squeeze was now definitely on the independent refiners who sold unbranded gasoline to many price marketers. Survival became a relevant question. "Several independent U.S. refiners shut down for extended turnarounds, or sharply curtailed what is obviously a losing proposition." [25]

The Great Price Restoration

The three years of price wars that had gripped much of the country after Gulftane was introduced came to an end generally on March 17, 1965. Texaco led a national price restoration by eliminating all special price concessions which had permitted its dealers to lower prices. Texaco's move was similar to the three national price restorations made by Gulf Oil Company from late 1962 to mid-1963. The big difference was that the Texaco restoration gained wide support and held.

What was the difference between Texaco's efforts and those of Gulf and the abortive move of Mobil? There are many theories and probably some truth in all of them. The profits of the two largest international oil companies, Texaco and Esso, had shown signs of weakening in the second half of 1964—a reflection of poorer international markets. The make up for the softening overseas markets, these companies had good reason to try to put the domestic market on a better paying basis. Other less important reasons probably included the increasing number of antitrust and below-cost suits that were being brought against the majors as smaller competitors found themselves forced into desperate financial straits. Furthermore, the Federal Trade Commission had cancelled four price suits against the majors and announced extensive hearings into the practice of marketing automotive gasoline. If conditions could be improved prior to the hearings less gory tales were likely to be presented than if price wars were still raging.

Reversal of the One-cent Differential Policy

While there were several reasons competitors would want Texaco's restoration to hold, this could not happen without some adjustments in the conflicting policies of both the majors and the price marketers. One of the big questions centered around the pricing of tanes relative to the price marketers' and majors' regular gasoline. Related to this was the question of the one-cent differential policy between the majors and price marketers on regular-grade gasoline. When Texaco announced the restoration it abandoned the one-cent differential in Wichita, Kansas, and Phillips did likewise in Kansas City and other markets. The general reversal of the majors' position regarding this practice was an important escape valve for many potentially explosive market situations. Reminding everyone what the one-cent differential policy could mean were markets such as Fort Worth, where one aggressive major was enforcing this policy. Almost a year after the general "restoration" Fort Worth still had not been restored to a normal price level.[26]

The response to the tanes of the price marketers was now generally to allow a one-cent differential. Supposedly, the majors would price their regular one cent above the tanes which would reestablish the traditional two-cent differential between the majors and price marketers. However, as it turned out the differentials were frequently much greater than two cents. Two years after the national price restoration by Texaco, the question of differentials seemed to be no longer particularly important. The majors appeared to be out to improve their realization from the domestic market and they tended to turn their backs on old issues that could potentially cost them big money. An example of this change is reported in an industry news brief.

> The traditional 2¢ spread between majors and private-branders seems to [be] out the window in many parts of the country.
>
> Present differentials range between 3¢ and 5¢, in key markets, says a national pricing executive. In Detroit and Chicago, it's a clean 4¢ to 5¢. In Denver, it's 3¢ across the board. The wider differentials have been in effect for some time and are not directly attributable to recent price activity.
>
> Spreads up to 7.5¢ and 10¢ gal. have been reported. The 10¢ spread occurred in Florida with the Phillips move.[27]

However, this should not be interpreted as meaning that market peace prevailed under such circumstances. Some price-sensitive areas in which prices were falling far below normal began to appear. Furthermore, in many markets where the price marketers were strong the two-cent differential was the prevalent policy. In other markets where the price marketers were less significant the wider differentials were found.

CONSEQUENCES OF TANE AND OTHER PRICE WARS

THERE SHOULD BE NO QUESTION of whether the prolonged price wars and below-cost selling of gasoline in the first half of the 1960's had an impact on the industry. Rather the question should be, what was the specific consequence of the abnormal behavior that took place in many regions and markets during this period of time? [28] As has been shown, one of the objectives at the beginning of the period was to stop the growth of price marketers, but the impact of the price warfare during this period was more far-reaching than simply its effect on the price marketers.

Varying Impact of Price Wars on Markets

BEFORE DISCUSSING the impact of the price wars on the various segments of the industry, the severity of the price wars in different areas of the country will be reviewed. As was previously discussed the introduction of tanes by Gulf and Sun became the catalyst for severe price disturbances in many areas. The unsettling effect of the tanes and the way they were reacted to by other majors and price marketers is illustrated by price changes in five volatile markets—Buffalo, Hartford, Minneapolis, Detroit, and Houston—shown in Figure 6-1. The method of analysis in general employed in Figure 6-1 and Tables 6-3 to 6-5 is that of determining the level of prices both before the introduction of Gulftane and after the return to normal following the Texaco price restoration of March 17, 1965. The pre-Gulftane and postrestoration periods are used as an indication of what price conditions were like during more normal periods of time. The period of the tane and widespread one-cent price wars falls in between the two base periods. (As was discussed there had been localized one-cent battles earlier in such areas as Birmingham, Norfolk, and Denver, and the West Coast.) The break in prices during approximately three years of market chaos in certain areas can be seen in the retail price of gasoline in the five major markets presented in Figure 6-1. The average price

FIGURE 6-1

BREAK IN PRICES IN CERTAIN MARKETS HIT PARTICULARLY HARD BY PRICE WARS FROM 1962 TO 1964

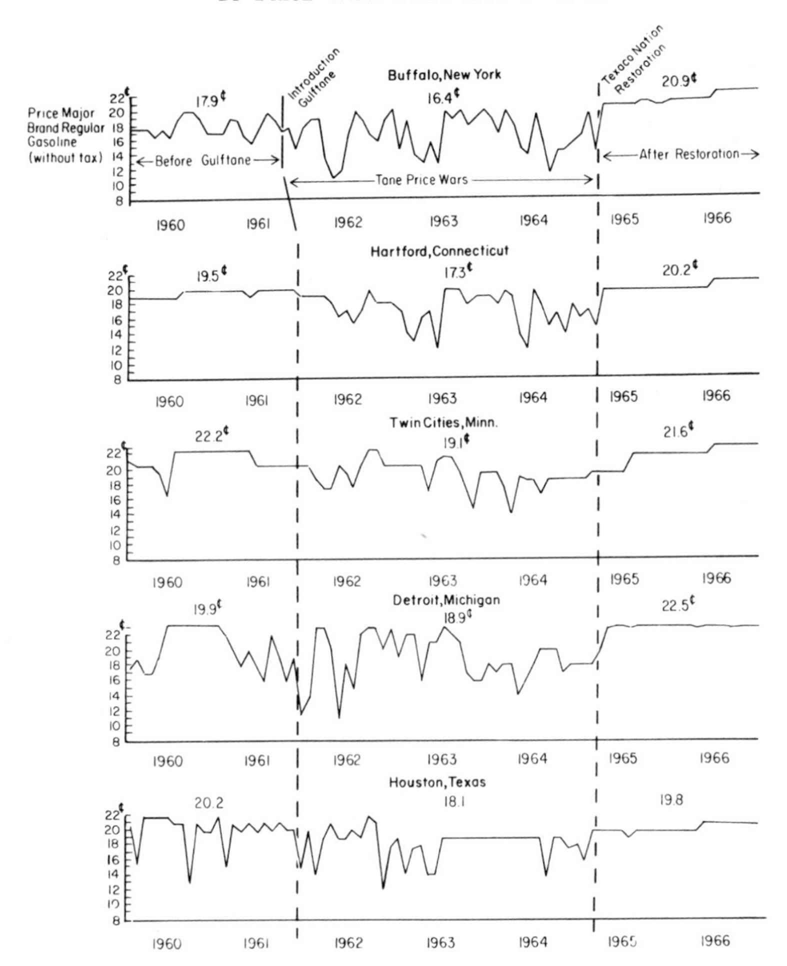

Source: Data derived from monthly price reports in the Platt's Oilgram Price Service.

of gasoline fell by several cents per gallon during the period of the tane and widespread one-cent differential price wars. The average difference in the retail price for these five markets between the normal and price-war periods is shown in Table 6-3. For the five markets the price of regular gasoline was from 1.9¢ to 3.0¢ below normal.

TABLE 6-3

DEPRESSION OF PRICES IN CERTAIN MARKETS
PARTICULARLY HARD HIT BY PRICE WARS

Markets	Average Price Before Price Wars (a)	Average Price After Price Wars (b)	Average Price During Normal Market Periods (c) (a + b ÷ 2)	Average Price During Price War Periods (d)	Lower Price During Price War Periods (e) (c − d)
Buffalo, N.Y.	17.9¢	20.9¢	19.4¢	16.4¢	3.0¢
Hartford, Conn.	19.5	20.2	19.8	17.3	2.5
Twin Cities, Minn.	22.2	21.6	21.9	19.1	2.8
Detroit, Mich.	19.9	22.5	21.2	18.9	2.3
Houston, Tex.	20.2	19.8	20.0	18.1	1.9

The data presented in Figure 6-1 also illustrate the nature of price changes in those areas experiencing frequent price wars. Price levels were not gradually changed, but rather were quickly driven down below the price marketers' cost as the battle over differentials went on. Often after a few weeks at cost, or below, prices would be restored. If competitors were willing to accept the terms of peace (typically a one-cent differential between the majors and the price marketers) then prices following a restoration would hold. Otherwise, prices would soon start working their way down again; this was the prevailing condition in certain markets. In markets experiencing long-run price problems the fluctuation in prices began to be known as the price "yo-yo."

Another characteristic of the severe price wars of the early 1960's was their varying impact on different markets and areas of the country. To determine which cities and areas were particularly subjected to the tane and one-cent differential price wars an analysis was made in which the average retail price of gasoline in the fifty-five cities traditionally considered indices to the market for the so-called normal years (1961, 1962, 1965, and 1966) was compared with prices during the price-war years (1962, 1963, and 1964). The results of this analysis are presented in Table 6-4. The data rather clearly show the uneven impact of the price-war years on these markets. For example, the prices in New York City (city no. 8) were hardly affected by the price-war period and averaged 98.6 percent of the so-called normal

TABLE 6-4

RELATIVE CHANGE IN PRICE FOR FIFTY-FIVE CITIES FROM "NORMAL" YEARS TO PRICE-WAR YEARS[1]

	Base Years		Price War Years			Base Years		Average Price During Base Years	Average Prices Relative to Base for Price War Years				Average Price Relative to Base for All Price War Years
	1960	1961	1962	1963	1964	1965	1966		1961	1962	1963	1964	
1. Portland, Me.	18.07	17.98	18.32	18.32	18.65	19.57	20.48	19.02		96.3%	96.3%	98.1%	96.9%
2. Manchester, N.H.	19.07	19.65	17.98	17.73	18.07	19.65	20.48	19.71		91.2	90.0	91.7	91.0
3. Burlington, Vt.	22.90	23.40	22.40	21.15	20.57	22.15	22.82	22.82		98.2	92.7	90.1	93.7
4. Boston, Mass.	17.48	18.65	17.57	17.48	18.07	19.07	19.98	18.80		93.4	93.0	96.1	94.2
5. Providence, R.I.	16.48	17.48	15.73	15.82	16.90	18.32	19.48	17.94		87.7	88.2	94.2	90.0
6. Hartford, Conn.	19.32	19.73	17.73	16.98	16.90	18.90	20.40	19.59		90.5	86.7	86.3	87.8
7. Buffalo, N.Y.	17.92	17.48[2]	16.46	16.77	15.77	19.31	21.40	19.03		86.5	88.1	82.9	85.8
8. New York, N.Y.	21.40	20.90	20.90	20.90	20.90	20.90	21.60	21.20		98.6	98.6	98.6	98.6
9. Newark, N.J.	19.32	19.90	19.40	19.40	19.90	19.90	21.48	20.15		96.3	96.3	98.8	97.1
10. Philadelphia, Pa.	19.48	19.40	17.65	17.65	17.65	18.32	20.32	19.38		91.1	91.1	91.1	91.1
11. Wilmington, Del.	20.07	19.57	19.40	17.82	19.32	19.32	20.48	19.86		97.7	89.7	97.3	94.9
12. Baltimore, Md.	21.65	20.98	18.57	18.98	19.24	19.57	21.57	20.94		88.7	90.6	91.9	90.4
13. Washington, D.C.	20.32	20.48	20.40	19.73	20.23	20.40	21.48	20.67		98.7	95.4	97.9	97.3
14. Charleston, W.V.	20.32	20.73	20.15	19.90	19.48	20.65	20.90	20.65		97.6	96.4	94.3	96.1
15. Norfolk, Va.	20.73	21.40[2]	17.84	19.90	19.90	19.73	20.57	20.61		86.6	96.6	96.6	93.3
16. Charlotte, N.C.	20.03	20.40[2]	19.98	19.57	19.65	20.57	21.48	20.62		96.9	94.9	95.3	95.7
17. Charleston, S.C.	19.94	20.90	20.90	20.90	21.32	20.90	21.48	20.80		100.5	100.5	102.5	101.2
18. Atlanta, Ga.	21.37	20.70	20.74	22.12	21.70	21.70	22.20	21.49		96.5	102.9	101.0	100.1
19. Jacksonville, Fla.	20.40	20.90	20.82	21.90	21.90	21.90	21.73	21.23		98.1	103.2	103.2	101.5
20. Birmingham, Ala.	19.23	19.65	20.63	19.73	20.78	21.73	20.48	20.27		101.8	97.3	102.5	100.5
21. Jackson, Miss.	21.23	21.52	20.28	20.14	22.37	21.13	22.10	21.50		94.3	93.7	104.0	97.6
22. Memphis, Tenn.	18.65	19.48	17.15	19.82	18.98	20.15	22.15	20.11		85.3	98.6	94.4	92.8
23. Louisville, Ky.	21.80	21.23	21.07	19.82	20.15	21.90	22.40	21.83		96.5	90.8	92.3	93.2
24. Cleveland, O.	21.32	21.32	20.90	20.90	20.90	20.90	21.73	21.32		98.0	98.0	98.0	98.0
25. Cincinnati, O.	20.40	20.57	19.90	19.90	19.90	20.40	21.73	20.78		95.8	95.8	95.8	95.8
26. Indianapolis, Ind.	20.73	21.90[2]	19.03	22.32	20.73	22.40	23.32	22.09		86.1	101.0	93.8	93.6
27. Chicago, Ill.	22.90	21.40	21.73	22.57	20.40	23.73	24.23	23.06		94.2	97.9	88.5	93.5
28. Detroit, Mich.	20.93	18.90	19.49	19.57	17.73	21.38	22.88	21.02		92.7	93.1	84.3	90.0

TABLE 6-4 (continued)

	Base Years		Price War Years			Base Years		Average Price During Base Years	Average Prices Relative to Base for Price War Years				Average Price Relative to Base for All Price War Years
	1960	1961	1962	1963	1964	1965	1966		1961	1962	1963	1964	
29. Milwaukee, Wis.	20.99	20.82	21.73	21.07	18.98	21.65	21.07	21.13		102.8	99.7	89.8	97.4
30. Twin Cities, Minn.	21.03	21.23	19.73	19.28	17.98	20.28	22.23	21.19		93.1	91.0	84.8	89.6
31. Fargo, N.D.	22.26	23.36	22.28	21.23	21.98	22.90	22.07	22.65		98.4	93.7	97.0	96.4
32. Huron, N.D.	24.07	22.98	22.82	23.57	20.40	20.90	20.65	22.15		103.0	106.4	92.1	100.3
33. Omaha, Neb.	21.52	21.48	21.23	21.48	18.82	20.78	22.82	21.65		98.1	99.2	86.9	94.7
34. Des Moines, Iowa	21.57	21.05	19.77	20.01	17.98	20.15	22.23	21.25		93.0	94.2	84.6	90.6
35. St. Louis, Mo.	18.65	18.90	18.94	19.57	18.15	19.73	23.23	20.13		94.1	97.2	90.2	93.8
36. Wichita, Kan.	20.15	22.75	20.65	16.40	18.32	19.82	19.73	20.61		100.2	79.6	88.9	89.6
37. Tulsa, Okla.	19.98	19.98	19.98	20.48	20.57	19.40	21.73	20.27		98.6	101.0	101.5	100.4
38. Little Rock, Ark.	18.98	19.32	18.98	19.65	20.40	20.40	21.07	19.94		95.2	98.6	102.3	98.7
39. New Orleans, La.	18.07	18.06	18.90	18.90	18.90	18.90	19.65	18.67		101.2	101.2	101.2	101.2
40. Dallas, Tex.	20.23	17.90	18.52	17.28	19.48	18.96	20.42	19.38		95.6	89.2	100.5	95.1
41. Houston, Tex.	20.32	19.57	18.53	17.44	18.38	19.32	20.23	19.86		93.3	87.8	92.5	91.2
42. El Paso, Tex.	19.89	20.73	21.90	21.90	21.90	21.90	22.90	21.36		102.5	102.5	102.5	102.5
43. Albuquerque, N.M.	22.40	19.48	20.98	20.15	22.40	21.86	22.40	21.54		97.4	93.5	103.9	98.3
44. Denver, Colo.	23.40	21.40	20.02	21.73	18.90	19.65	21.86	21.58		92.8	73.8	87.6	93.3
45. Cheyenne, Wym.	25.98	24.48	23.32	23.40	24.65	23.07	20.82	23.59		98.8	99.2	104.5	100.8
46. Great Falls, Mont.	26.90	26.90	27.65	24.65	24.13	24.09	23.82	25.42		108.8	97.0	94.9	100.2
47. Boise, Idaho	27.42	25.98	26.07	24.98	24.82	24.65	25.07	25.78		101.1	96.9	96.3	98.1
48. Salt Lake, Ut.	20.90	21.73	23.08	21.64	21.57	21.65	20.90	21.30		108.4	101.6	101.3	103.8
49. Reno, Nev.	28.00	24.36	22.57	21.65	23.07	21.32	22.23	23.98		94.1	90.3	96.2	93.5
50. Phoenix, Ariz.	17.45	20.73	21.15	21.65	19.32	19.23	22.32	19.93		106.1	108.6	96.9	103.9
51. Los Angeles, Calif.	19.57	18.65	20.23	19.57	19.73	21.32	20.65	20.51[3]	90.9	98.6	95.4	96.2	95.3
52. San Francisco, Calif.	22.23	19.73	21.90	21.86	21.90	21.90	22.23	22.12[3]	89.2	99.0	98.8	99.0	96.5
53. Portland, Ore.	23.65	20.23	20.04	19.07	19.48	20.73	22.07	22.15[3]	91.3	90.5	86.1	88.9	89.2
54. Seattle, Wash.	19.07	20.73	20.48	19.07	17.82	19.15	20.40	19.54[3]	106.1	104.8	97.6	91.2	99.9
55. Spokane, Wash.	26.23	19.90	21.48	20.57	20.32	20.36	20.23	22.27[3]	89.4	96.4	92.4	91.2	92.4
Average U.S.	20.99	20.71	20.25	20.11	19.88	20.70	21.57	20.99					
%	100.0	98.7	(96.5)	(95.8)	(94.7)	98.7	102.8	100.0					

1. Compiled from monthly data presented by *Platt's Oilgram Price Service*.
2. The last four months of 1961 are combined with 1962 data since Gulftane was introduced earlier in these markets.
3. Data for 1961 not used in calculating base since one-cent differential war started approximately a year earlier in these markets.

price. In contrast, the nearby cities of Hartford (city no. 6) to the north and Baltimore (city no. 12) to the south experienced severely depressed prices that fell to respectively 87.8 percent and 90.4 percent of normal. The extent to which some cities seemed to be relatively immune to price wars while others were hit hard can also be visualized from Table 6-5 where cities with reasonably normal prices are grouped together and other cities with depressed prices are divided into three categories varying in severity of the depression. The difference in price conditions between markets and regions of the country will be woven into the analysis of the impact of the price wars on various members of the industry structure that follows.

A Price-war Profit Paradox

While gasoline prices were severely depressed in many markets for prolonged periods from 1961 to 1964, these adverse market conditions were frequently not reflected in the oil companies' earnings reports. During this three-year period of price turbulence the overall profit of the major oil companies actually improved (see Table 6-6). An analysis of the quarterly profit reports of the major oil companies leads to the following conclusions about the price-war profit paradox.

1. The price wars had widely differing impact on the performance of the major oil companies.
2. General profit improvement was not associated with gains in the profitable marketing of gasoline.

The differential impact of the price wars on the performance of individual companies is revealed by their earnings reports. The profit history for the major oil companies during the first half of 1962, which reflected the introduction of tanes in most markets, gives some perspective into what was happening (see Table 6-7). All twelve of the companies showing declining profits (nos. 16 to 27) had rather extensive operations in the Mid-continent and in the New England states, where subregular grades of gasoline were "shaking up markets" (see Table 6-5). Not surprisingly Gulf and Sun, the two majors that were aggressively pricing tanes against the price marketers had sharply depressed profits (nos. 19 and 22).

With Gulf's nationwide price restoration in September 1962 there

TABLE 6-5

DIFFERENTIAL IMPACT OF PRICE WARS ON THE FIFTY-FIVE MAJOR CITIES CONSIDERED MARKET INDICES

Levels of Retail Price Below Normal for Price War Years

Region of Country	3 9 cities (90% Base or Less)	2 13 cities (91–93% Base)	1 11 cities (94–96% Base)	Normal Prices 22 cities (97–103% Base)
New England	Providence, R.I. Hartford, Conn. Buffalo, N.Y.	Manchester, N.H. Burlington, Vt.	Portland, Maine Boston, Mass.	
Mid-Atlantic	Baltimore, Md.	Philadelphia, Pa. Norfolk, Va.	Wilmington, Del. Charleston, W.Va.	New York City, N.Y. Newark, N.J. Washington, D.C.
Southeast			Charlotte, N.C.	Charleston, S.C. Atlanta, Ga. Jacksonville, Fla. Birmingham, Ala. Jackson, Miss.
Mid-continent	Detroit, Mich. Twin Cities, Minn. Des Moines, Iowa Wichita, Kan.	Memphis, Tenn. Louisville, Ky. Indianapolis, Ind. Chicago, Ill. St. Louis, Mo.	Cincinnati, O. Fargo, N.D. Omaha, Neb.	Cleveland, O. Milwaukee, Wis. Huron, N.D.
Southwest		Houston, Tex. El Paso, Tex.	Dallas, Tex.	Tulsa, Okla. Little Rock, Ark. New Orleans, La. Albuquerque, N.M.
Near West		Denver, Colo. Reno, Nev.		Cheyenne, Wy. Great Falls, Mont. Boise, Idaho Salt Lake City, Utah Phoenix, Arizona
Far West	Portland, Ore.	Spokane, Wash.	Los Angeles, Calif. San Francisco, Calif.	Seattle, Wash.

TABLE 6-6

CHANGES IN PROFIT PERFORMANCE OF THE MAJOR VERTICALLY INTEGRATED OIL COMPANIES FOR THE GASOLINE PRICE-WAR YEARS OF 1961 TO 1964

	1961-1962	1962-1963	1963-1964
Standard Oil New Jersey:	11%	21%	3%
Texaco:	11	13	5
Gulf:	0	9	6
Socal:	7	3	7
Mobil:	15	12	8
American:	6	12	6
Shell:	12	14	10
Phillips:	−6	7	−1
Continental:	7	18	15
Cities Service:	10	10	9
Sun:	7	15	12
Sinclair:	32	32	−6
Atlantic:	0	−6	7
Union:	24	16	21
Marathon:	−3	30	23
Sunray DX:	−10	16	−16
Sohio:	−5	40	29
Tidewater:	17	21	−10
Pure:	−4	3	6
Richfield:	22	−9	−23
Weighted Average:	8%	12%	6%

Source: Data from National Petroleum News Factbook.

was considerable market improvement in certain areas—particularly in the mid-Atlantic states and in the Southeast. In many markets prices remained strong for more than a year, or into the fourth quarter of 1963. However, markets in the Mid-continent and in the New England coastal areas (including Buffalo) remained rather price turbulent (see Tables 6-4 and 6-5).[29]

Profit Improvement Not From Marketing

MANY OF the statements associated with earnings reports during the price-war years of 1962 to 1964 have indicated that the general profit improvement of these years was not from marketing. Consider for example a number of comments made with the improved earnings reports for the first half of 1964.

> In general, all companies have made less from marketing so far this year than they did last year at the same time. . . .

TABLE 6-7

CHANGES IN OIL COMPANY EARNINGS FOR FIRST HALF OF 1962

	1962 (add 000)	1961 (add 000)	% Change 1961 to 1962
1. Signal Oil and Gas:	10,620	5,927	+79.2
2. Richfield:	12,546	9,180	+36.7
3. Union:	20,576	16,771	+22.7
4. Hess Oil:	4,191	3,587	+16.9
5. Imperial:	33,424	29,091	+14.9
6. Tidewater:	16,598	14,718	+12.8
7. British American:	14,870	13,267	+12.1
8. Standard Oil of New Jersey:	425,000	385,000	+10.4
9. Texaco:	220,268	200,151	+10.1
10. Shell:	74,937	69,683	+7.5
11. Socony Mobil:	117,400	108,800	+7.9
12. Standard Oil of California:	153,903	142,828	+7.8
13. Continental:	31,400	29,900	+5.0
14. Pure Oil:	11,363	10,874	+4.5
15. Atlantic:	22,444	21,649	+3.7
16. Skelly:	10,841	11,813	−8.2
17. Phillips:	50,130	54,902	−8.7
18. Standard Oil of Indiana:	67,658	74,177	−8.8
19. Gulf:	151,000	169,701	−11.0
20. Sohio:	10,112	11,640	−13.1
21. Cities Service:	25,066	29,511	−15.1
22. Sun Oil:	19,664	24,048	−18.2
23. Sunray DX:	17,889	21,916	−18.4
24. Marathon (Ohio):	15,541	19,300	−19.5
25. Sinclair:	18,657	24,654	−24.3
26. Apco:	1,004	1,702	−41.1
27. Clark:	(572)	17	—
Total 27 companies:	1,556,530	1,504,807	+3.4

Source: National Petroleum News Bulletin (*August 13, 1962*), *p. 1.*

A national major with a glowing first-half report says it made "substantially less" from manufacturing, marketing, and transportation than it did last year. Production and a bright international picture make it look good.

It isn't possible to make more money in marketing, . . . says a disgruntled marketing V.P.

Again in general, the internationals tended to do better than others. Some of the foreign-crude liftings have been lucrative.

Unit realizations are so poor, one marketer says, "Return on gasoline makes you sick." [30]

Gulf Oil's chairman, William K. Whiteford, said in reporting record earnings for 1964 that "Gasoline prices were terrible in 1964 both in the U.S. and abroad, but we're making good money off new crude-oil production." [31] Finally, the *NPN Bulletin* in recapping the 1964 earnings reports said, "Profits for many major companies hit record levels in 1964, but most of them said it was despite depressed product prices." [32]

In summary, the earnings reports of the major oil companies tended to show two basic reasons for the "price-war profit paradox." (1) The price wars were frequently regional in nature—hardest hit were the Mid-continent (Midwest and Southwest) and the New England coastal markets. Those oil companies with widespread domestic operations and international diversity were able to average their earnings and to do quite well. For example, Chevron reported that during mid-1963 some 60 percent of its 261 trade areas were at "normal" prices, 7 or 8 percent of its markets were above normal, and the remainder were depressed.[33] Furthermore, Chevron, being an international oil company, was able to improve its profits by the exceptional returns from its extensive foreign operations as were the other international majors. (2) The earnings reports also covered up the depressed market prices because of profits from other activities such as improvements in crude-oil production and diversified activities such as petro-chemical operations.

Effects of Price Wars on Major Oil Companies

While the oil industry in general passed through the turbulent 1962 to 1964 period relatively unscathed, many of the regionally concentrated majors did not. The tane and differential price wars of the early 1960's forced the merger of many of the smaller majors with regionally concentrated operations, and especially those without much crude oil. The rationale behind the mergers can in part be assessed from studying the financial performances of the major oil companies for the period 1960 to 1964 which are shown in Table 6-8. The international and domestic oil companies are ranked according to their five-year compounded growth rate in terms of sales and profit. For example, Standard Oil of Ohio ranked the highest for the five-year

TABLE 6-8

FIVE-YEAR YARDSTICK OF PERFORMANCE (1960 TO 1964) RELATED TO MERGERS

	Sales	Earnings	Rank
International:			
Texaco:	6.5%	9.3%	1
Mobil:	7.8	7.9	2
Standard Oil of New Jersey:	5.7	8.4	3
Gulf:	1.7	4.5	4
Standard Oil of California:	6.2	3.3	4
			Standard Oil of Kentucky
Domestic:			
Standard Oil of Ohio:	4.6	9.4	1
Union—California:	3.5	11.1	2
Cities Service:	3.6	6.2	3
Shell:	4.0	5.6	3
Atlantic Ref.:	2.6	9.3	4
Marathon:	9.1	4.6	4
Continental:	7.7	4.4	5
Sun:	2.5	8.9	5
Standard Oil of Indiana:	2.7	5.9	6
Tidewater:	3.2	4.2	7
Phillips:	2.6	3.2	8
Sunray DX:	3.6	decline	9
Pure:	3.9	decline	10
Richfield:	2.9	1.6	10
Sinclair:	decline	2.5	11

Source: "*Oil*," Forbes (*17th Annual Report on American Industry*), (*January 1, 1965*), *p. 26.*

1. Citgo sold its nine midwestern state markets to Gulf.
2. Phillips purchased Tidewater's West Coast marketing and refining operation.
3. The overlapping Sinclair and Atlantic operations on the East Coast were sold to British Petroleum which later merged with Sohio.
4. Standard Oil of California took over Standard Oil of Kentucky in 1961.

period while Sinclair ranked the lowest. During the five-year period, Sinclair's sales actually declined and its profit performance was far below average.

The arrows used in Table 6-8 indicate that five of the six poorest performers were merged into companies with better overall performance records. Those companies or divisions of companies absorbed were in most cases suffering from the effect of regional price wars. As previously discussed Richfield, a seven-state West Coast operation, and Tidewater's West Coast division were suffering from the ill effects of Shell's one-cent West Coast policy which led to recurring price wars (see Figure 6-2 for a description of assets sold). Sunray DX's seventeen-state operation and Citgo's nine-state midwestern operation suffered greatly as a result of the Mid-continent price wars of the early 1960's. Both Pure and Sinclair were heavily concentrated in the Mid-continent and also suffered as a result of the price wars (see Figure 6-2 for marketing territories involved). *Forbes* magazine, commenting on the reasoning behind the merger, noted: "In 1965 Union acquired Pure Oil, a sickly midwestern and southern states refiner and marketer that was being nibbled to death by price wars in its area." [34] Phillips and American Oil also reported poor earnings for the five-year period 1960 to 1964. Both were relatively strong in the price-torn Mid-continent area and they made it through the early 1960's without merging because of their sheer size and the diversity of their operations.

The major-brand dealers and jobbers operating in the areas hard hit by price wars also in general suffered. Typically, they absorbed 50 percent of every one-cent reduction in the retail price down to a stop-out margin. The net result of lengthy price wars was reduced dealer and jobber profits. (For most dealers and jobbers, volume does not increase enough to compensate for the reduction in margin.) The consequence was financial hardship and increased dealer turnover.

Effects of Price Wars on the Independent Refineries

The effects of the price wars of the early 1960's (and those of the latter 1950's) on the independents was analogous to what happened to the regionally concentrated majors. Under the severe financial pressure of the price wars, many independents sold out to larger, more diversified firms that were able to withstand price wars. Frequently, those that came through the period were the independents that had sufficient regional diversity and other sources of profit to help them through the extended periods of depressed regional gasoline prices.

As was shown from the analysis of the impact of the price wars on certain cities in Tables 6-4 and 6-5 and the discussion of majors particularly hurt by the price wars, the Mid-continent area was one of the most hard hit regions of the country. It was also the primary area in the country where independent refiners were located. There are very few independent refiners on the East Coast and only several small independent refiners on the West Coast.

FIGURE 6-2

Characteristics of Companies Taken Over by Mergers

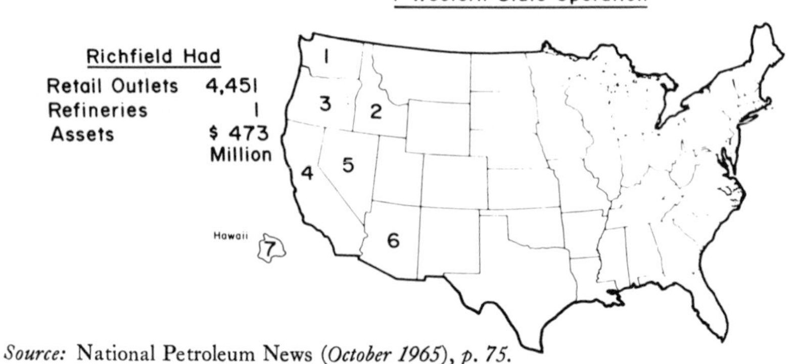

Source: National Petroleum News (*October 1965*), *p. 75.*

Tidewater Sold to Phillips
4 Primary States of West Coast Operation

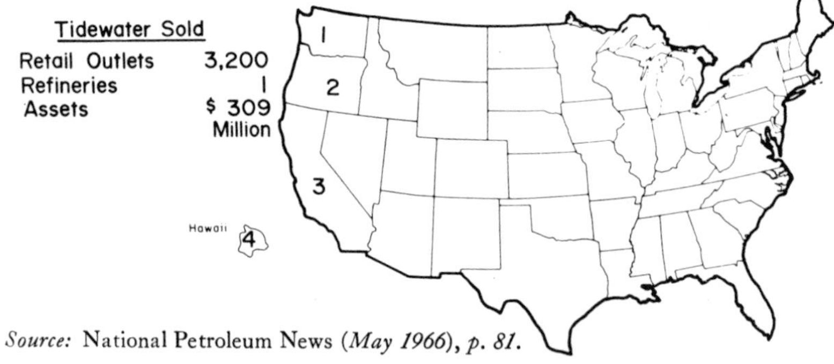

Source: National Petroleum News (*May 1966*), *p. 81.*

Pure Merged into Union
24 State Operation

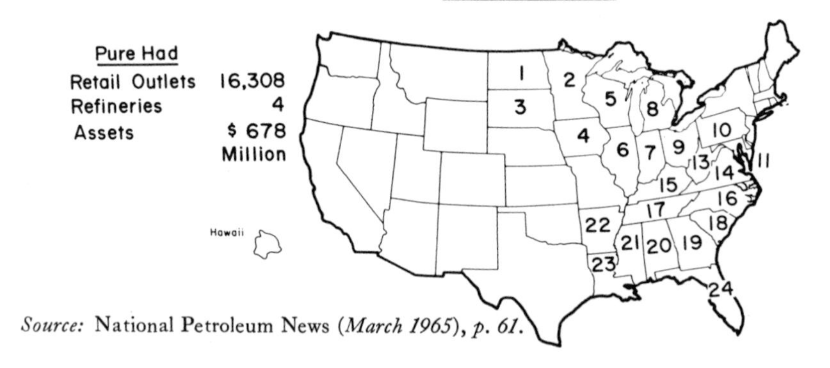

Source: National Petroleum News (*March 1965*), *p. 61.*

FIGURE 6-2 (continued)

CHARACTERISTICS OF COMPANIES TAKEN OVER BY MERGERS

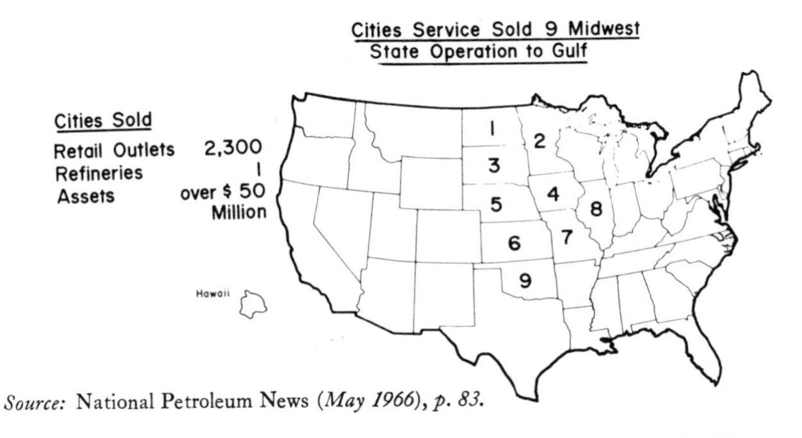

Source: National Petroleum News (*May 1966*), *p. 83.*

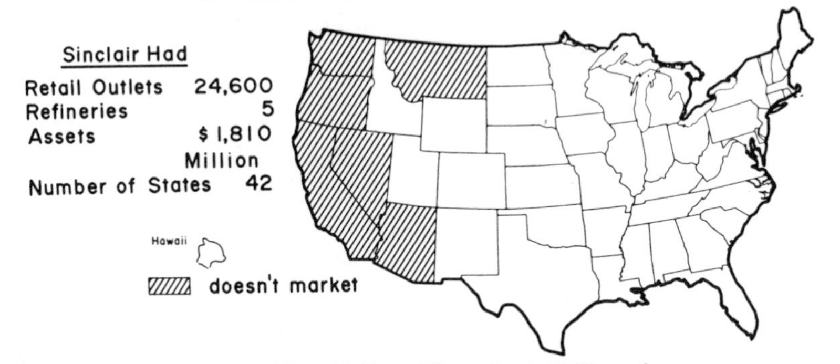

Source: National Petroleum News Bulletin (*November 4, 1968*), *p. 1.*

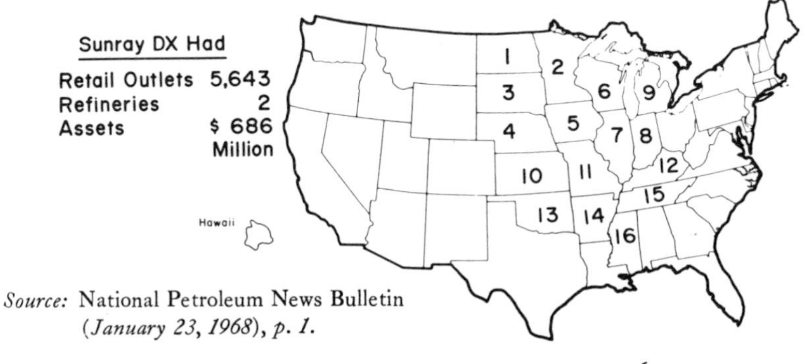

Source: National Petroleum News Bulletin (*January 23, 1968*), *p. 1.*

The financial squeeze that the independent segment of the industry began to experience late in 1961 led to the formation of a coalition of many of the significant mid-continental refiners. Early in 1963, seventeen independent refiners formed an organization called MIRA—Mid-continent Independent Refiners Association—whose central purpose was to present a unified front of independents to protest the destructive nature of competition which they felt would eventually destroy the independent segment of the industry. For such a group to have been formed says something about the severity of the problems the independents faced. In many cases the independents were arch competitors and it took a crisis of major proportions to get them to cooperate in the first place, and then, to present a unified front against what was to them destructive developments in their industry.

A majority of the significant independent refiners-marketers in the Mid-continent joined MIRA.[35] They included:

American Petrofina Co.	Dallas, Texas
Apco Oil Co.	Oklahoma City, Oklahoma
Bell Oil and Gas Co.	Tulsa, Oklahoma
Champlin Oil and Refining Co.	Fort Worth, Texas
Clark Oil and Refining Co.	Milwaukee, Wisconsin
Consumers Cooperative Association	Kansas City, Missouri
Derby Refining Co.	Wichita, Kansas
Farmers Union Central Exchange, Inc.	Oklahoma City, Oklahoma
Kerr-McGee Oil Industries, Inc.	Oklahoma City, Oklahoma
Midland Cooperative, Inc.	Minneapolis, Minnesota
National Cooperative Refining Assoc.	McPherson, Kansas
Northwestern Refining Co.	St. Paul, Minnesota
Premier Oil Refining Co.	Houston, Texas
Texas City Refining, Inc.	Texas City, Texas
Vickers Petroleum Corp.	Wichita, Kansas

However, within two years of its founding, five of the original seventeen companies forming MIRA had merged or sold out, and others had shut down their refining operations. The failure of the independents was associated with the financial bind that they found themselves in—the crude-oil price was relatively fixed and the price wars (tane and one-cent differential) were driving down the average realization from refining and marketing. The nature of the squeeze on independent refining margins can be observed from Table 6-9.

TABLE 6-9

SQUEEZE ON REFINING MARGINS OF THE INDEPENDENT REFINERIES

	Average for 4 Products $/Bbl	Average Price of Crude Barrel	Refinery Margin
1956	$3.91	$2.89	$1.02
1957	4.14	3.18	.96
1958	3.79	3.15	.94
1959	3.83	3.09	.74
1960	3.84	3.10	.74
1961	3.87	3.10	.77
1962	3.82	3.08	.74
1963	3.76	3.08	.68
1964	3.69	3.07	.62

Source: U.S., Congress, House, Select Committee on Small Business, "The Impact Upon Small Business of Dual Distribution and Related Vertical Integration," 89th Cong., 1st sess. (Washington, D.C.: U.S. Government Printing Office), p. 98.

The seriousness of the problem faced by the independent refiners was presented by the editor of *Platt's Oilgram Price Service* as follows:

> The year 1964 will go down as the one in which either the independent refiner or the price of crude couldn't quite make it. Right now the odds look to be against the independent.
> Last week, the Gulf Coast cargo market went into a violent convulsion.
> The latest cuts put gasoline prices at the Gulf back 18 years. Not since 1947 have gasoline prices been as low as 9.75¢ for 90 octane in the Gulf, and in those days East Texas crude was an average of $2.00 bbl. (against today's $3.10).
> The reductions, in effect, put the refiner's margin at the Gulf down to nothing. There is no gross margin whatsoever at today's wholesale prices. It was roughly only 30¢ bbl. before the present cuts, and last week's cuts certainly extinguished it.
> Much the same sorry story was being unfolded in the mid-continent. Last week, gasoline fell to 9.25¢ at wholesale. This is a level that hasn't obtained for 15 years—since the depression year of 1949. The mid-summer margin of Oklahoma refiners was estimated (at wholesale) to be only about 45¢ bbl. At present prices, the return on bulk sales does not cover the cost of crude and refining.[36]

Two months later the dismal prognosis for the independents was amplified:

WHAT'S GOING TO HAPPEN TO THE INDEPENDENT?

What is the future of the independent? Can he survive in today's business climate? Or is it a prerequisite of financial health to be big and integrated?
The question is germane, for the recent weeks have seen several independent U.S. refiners shut down for extended turnarounds, or sharply curtail what is obviously a losing operation.[37]

The exit of independent refiners during this period is illustrated by three significant mergers. On March 31, 1963, Delhi-Taylor Oil Corporation, a supplier of unbranded gasoline, was sold to Hess Oil and Chemical Company for $25 million cash. Delhi was a large operation itself with a refining capacity of 55,000 barrels of crude oil per day, 18 terminals, 106 service stations; it owned 40 percent of Billups Eastern Petroleum Company, which had 170 stations. However, Hess's sales were three times that of Delhi-Taylor and Hess was located in the mid-Atlantic area where prices were not nearly so depressed as in those areas in which Delhi-Taylor sales were concentrated.[38]

Another merger of significant independent refiners was the purchase of Cosden Petroleum for $75 million by American Petrofina. Both had large sales volumes in the Mid-continent. Aspects of this merger and others that American Petrofina was involved in are shown in Table 6-10. As the table indicates American Petrofina's growth was

TABLE 6-10

CHARACTERISTICS OF COMPANIES ACQUIRED BY AMERICAN PETROFINA

Company Acquired	Approximate Date of Takeover	Refining Capacity	Number of Jobbers	Number of Stations
Panhandle Oil:	Early 1957	7,500	95	250
American Liberty:	Early 1957	16,000	52	400
Atlas:	Early 1958	20,000	175	560
Cosden:	Early 1963	37,000	154	765
Total:		80,500	476	1,975

to a very great extent the result of the merging of independent refiners.[39] American Petrofina's strategy was one of balance, of trying to expand controlled outlets under the Fina brand to absorb most of its refiners' runs, and of increasing its production of crude oil.

Some of the other independents that survived the price-war period, such as Murphy Oil and Champlin, were also companies which had grown as a result of the merger of independents. Such mergers strengthened the surviving companies by providing (1) wider market coverage and market diversification which reduced the effects of localized and regional price wars; (2) increased sales through branded outlets which reduced the companies dependence on the depressed unbranded markets; and (3) production of crude oil which tapped a more profitable aspect of the industry's activity.

The final example of a sellout by an independent refiner was that of Premier Oil Refining early in 1964. Premier was one of the last of the original independent refiners and its departure marked the end of an era. Premier sold its four small refineries and 500 branded service-station accounts located in Texas, Louisiana, and Oklahoma to Sunray DX and Vickers Petroleum—50 percent to each.

Effects of Price Wars on Independent Price Marketers

The price wars not only had a severe effect on the independent refiners, but also on the independent marketers. Their most extensive market positions were in the Mid-continent (Midwest and Southwest) and the Southeast. These were the areas in which there was generally good assurance of an unbranded-gasoline supply at reasonable prices which was one of the prerequisites for the growth of the private branders.

As was shown in Table 6-5 many parts of the Midwest were particularly hard hit by the price wars which very definitely had a deteriorating effect on the independent marketers selling unbranded gasoline. In addition, the price marketers were penetrating certain other markets where ocean or inland terminals could obtain and were making available unbranded gasoline to private branders. The extremely low prices of gasoline shown in Tables 6-3 and 6-4 for the principal markets in New England—Boston, Providence, Hartford, and Springfield, plus Buffalo—were associated with the penetration of these markets by the private branders. In Boston, Gibbs Oil Company was a source of supply for private branders such as Tulsa, Bay State, and Tri S. Sources of supply of unbranded gasoline in other major New

England markets included General Oil in Providence, Murray Oil in Hartford, and Roberts Oil in Springfield. In Buffalo which was very hard hit by price wars, Frontier Refining (Ashland) and Aurora Refining (Marathon) were major suppliers of private branders. Similarly, prices were quite depressed in certain mid-Atlantic markets where private branders had become a factor. This was the case in Philadelphia and Baltimore where prices were depressed two or three levels (see Table 6-5); while in New York City and Washington, D.C., where price marketers had basically no position, prices remained strong.

One of the principal consequences of the concentrated price wars was that in many markets they brought to an end the era of private-brand operators who ran as few as one to upwards of twenty-five to fifty or even a hundred stations in and around a given metropolitan area. The prices were so volatile and deeply cut in these markets to often make it impossible for concentrated private branders to operate profitably and survive. As a result many localized private branders either sold out to majors or integrated independents. Others gained price protection by contracting to buy gasoline from a given supplier* or else by converting to major brands (Phillips, Texaco, and Mobil), major secondary brands (Seaside, Harbor, and Rocket), or independent refiners' brands (Apco, Fina, and Deep Rock).

In California and the West Coast where Shell initiated its one-cent policy in early 1961 large numbers of independent marketers disappeared from the scene in a variety of ways. The new majors on the West Coast—Esso, American, and Conoco—purchased hundreds of independents ranging from single stations to private branders with several hundred units.[40] Independents who did not sell out, but could not withstand the uneconomical level of prices, converted to major brands (frequently to Texaco) with price support.[41] Other West Coast private branders gained price support by changing to secondary brands of major oil companies such as Atlantic Richfield's Rocket brand.

The pattern of exit by the price marketers in the Mid-continent was quite similar to what occurred on the West Coast. The changes in the market structure in San Antonio, Texas, illustrate what happened as a result of the prolonged price wars in many markets where price mar-

* Price protection is a method by which a supplier temporarily reduces the prices at which he sells gasoline to distributors and service-station operators during price-war periods. Price protection is discussed at length in Chapter 7.

keters were significant. Before Gulftane, San Antonio was a big independent market. With approximately 20 percent of the stations, the independents were estimated to have been doing 45 percent of the gasoline business by selling for two cents or more under the majors' prices. The independents found themselves in real trouble when prices tumbled to far below their cost levels for extended periods of time. The effect of the price wars on many private branders was clear and predictable. As the *National Petroleum News* said in the early days of Gulftane, "It will take plenty of skill, guts and money to weather the price chaos surrounding the new gasoline grades." [42] While many private branders had the "skills" and the "guts," they were deficient in what the majors had more of—"money," financial staying power. Basically there was no way for the local and regional private branders to withstand the prolonged levels of below-cost selling associated with the introduction of Gulftane. Private branders in San Antonio closed stations and sold out, or became branded operators for majors such as Phillips, or converted to integrated independent refiner brands such as Fina and Shamrock.[43]

To survive many one-time private branders sacrificed their independence and "contracted" to buy from majors such as Phillips or integrated independent refiners such as Shamrock. In the process the one-time private branders typically withdrew from buying lower-price unbranded gasoline, but gained the all important price support which they had to have to withstand the price wars and prolonged periods of below-cost selling.[44] By becoming part of a larger system of business operations certain one-time private branders were able to gain at a cost (higher normal price and loss of freedom of action) the advantages of widespread market operations and other diversity of the bigger business units with which they were becoming associated.

In addition to private branders "contractually integrating" to gain "price protection," there were a large number of mergers of private branders into integrated companies which were better able to withstand the uneconomic conditions existing in many markets. For example, in 1960 Spur, one of the biggest private branders with over 300 stations located in twenty southern states, was merged into the Murphy Oil Company. A year later Ingram with approximately 200 branded outlets was also merged into Murphy. Another large private brander, Direct Oil, with 132 stations concentrated in eleven southeastern states, sold to Tenneco Oil in 1962. In January 1963 partial ownership of the 170-station Billups Eastern private-brand operation centered in the Southeast passed to Hess Oil and Chemical and in 1965 Hess gained complete control of the chain. These three mergers (Spur, Direct, and Billups Eastern) plus the purchase of Billups Western (by Signal Oil and Gas) and Kayo (by Continental) eliminated the most significant private-brand chain operations in the South and Southeast. This was basically the same thing that happened on the West Coast.

Many of the private branders who made it through the turbulent period of the early 1960's without price support had widely scattered operations. These include operations such as Hudson, Martin, Site, Star, and Imperial Refineries which typically operated 100 to 200 high volume stations widely dispersed over fifteen to thirty states. By dispersing they were able to gain protection against regional price wars in much the same manner as the major oil companies with tens of thousands of stations scattered throughout forty to fifty states did. Their spread-out approach led them to be described by some as "one station per market operators."

As a result of the price wars of the early 1960's many of the surviving private branders obtained even greater regional diversity. Both Hudson and Martin pushed into the West Coast from their headquarters in the Midwest. Star's approach was not only to spread out, but to lessen its concentration in markets hard hit by severe price wars. In St. Louis, which was Star's headquarters, Star sold its twenty-three stations to Phillips.

Summary of Impact of Price Wars on the Independent Segment of Industry

In summary, the effects of the price wars on the independents varied according to the three principal types of independents—private branders, refiner-marketers, and small integrated independents (see Table 6-11).

Starting first with the private branders the direction of change is clear. Large numbers of private branders sold out to major oil companies or to integrated independents. As the *National Petroleum News* said about the private branders, "There have been hundreds of deals, relatively few significant starts, and a net loss in private brand population." [45]

The effect of the price wars on the independent refiners and on the refiner-marketers was that many of them sold out to the integrated independents or major oil companies. The independent refiners and refiner-marketers were caught in the squeeze between falling market and wholesale prices for gasoline and the fixed prices of crude oil.

TABLE 6-11

CLASSIFICATION OF INDEPENDENTS

Functions	Integrated Independent	Refiner-Marketer	Independent Refinery	Private Brander
Marketing	X	X		X
Refining	X	X	X	
Production	X			

The only segment of the independent industry that seemed to grow as a result of the years of turbulent prices was the integrated independents. They owned much of the crude oil (20 to 70 percent) used by their refineries. As a result the integrated independents were less pressured in the squeeze of falling wholesale and market prices for gasoline and fixed prices for crude oil than were the refiner-marketers and independent refiners. For example, the president of Derby Refining, the refining and marketing arm of Colorado Interstate Corporation, testified before the Federal Trade Commission in 1965 that his refining-marketing operation was losing money "every month" (referring to much of the three years of depressed prices in his area). Were it not for the large crude-oil production and profits of the parent company Colorado Interstate Corporation, Derby would probably have gone the way of other independent refiners and refiner-marketers. In addition, the integrated independents frequently, as a result of mergers, had wide market coverage which gave them better opportunities to average out the bad and good markets. In fact, the branded representation of the integrated independents grew during this period of time since they were able to offer the umbrella of price protection to local and regional private branders who could not withstand the depressed price conditions of the early 1960's. Rather than being forced out of business many private branders contracted to sell the integrated independents' brands such as Fina, Shamrock, and Champlin. However, as a result of converting to an independent-refiner brand with price protection, the one-time private branders were generally less price aggressive. With the refiners' brand they were more inclined to accept for a variety of reasons a one-cent differential than they had been when they were private branders.

7 / *Price Protection Programs*

THE SELECTIVE PRICE WARS that have been discussed in the previous two chapters were financed by the price-subsidy programs of the major oil companies. The principal method by which the oil companies subsidize price cutting is through their price-protection programs (sometimes referred to as price assistance or price allowance). What price protection amounts to is the selective reduction of prices to gasoline purchasers at major-brand stations located in certain regions of the country, cities, or in particular parts of cities.

Price protection plays the very important role of defending and protecting the nonprice competitive strategy of the major oil companies. As previously explained (see Chapter 2, pp. 11–16) the majors' nonprice method of operation depends upon the maintenance of stable and relatively high prices. What price protection does is allow the major oil companies to combine in certain areas low price with their costly nonprice tools of competition to squelch, discipline, and otherwise reduce the effectiveness of companies that have chosen to compete on a price basis. While price protection is sometimes used by the major oil companies to police one another, so that no major gets away with price cutting for long, its more frequent target is the price marketers who specialize in selling gasoline on a price-discounted formula. It is not in markets like New York City, Washington, D.C., San Francisco, Cincinnati, and Hawaii where the majors dominate the markets that price protection is commonly employed. In these areas prices are stable and price wars financed by price protection are infrequent. In contrast in markets like Indianapolis, St. Louis, Kansas City, Denver, and Phoenix where the majors are not dominant price protection finances periodic price wars.

Price wars fed by price protection are an abnormal type of competition. The financial bludgeon employed in such price wars is made possible by drawing upon resources generated from the less competitive areas and activities where prices and profits are good. In this sense price protection is a mechanism of cross-market, or vertical, subsidization with financial power being used to regulate the nature of competition in the market place.

Mechanism of Price Protection

Many aspects of the major oil companies' method of selling gasoline are structured to minimize the likelihood of branded dealers reducing prices on their own. This includes major oil company ownership of real estate, short-term leases, and close supervision. In addition, dealer margins are kept relatively narrow so that the vast majority of dealers would find it difficult to reduce prices on their own.

With the major oil company system designed to make it generally unprofitable for dealers to reduce their own prices, the majors are in a position to centrally control the pricing of their dealers. The technique used to change the prices at which their dealers sell gasoline is the granting of price protection. To appreciate how price protection works consider the following example. While somewhat on the high side, assume that the major-brand dealer has a six-cent operating margin (after deducting rent). If the dealer lowered his price by two to three cents he would be cutting his operating margin by 33 to 50 percent. Since the vast majority of dealers is hard pressed as it is because of the small volume of business they do, few are prepared to lower their prices and reduce their own margins. However, if a dealer receives price protection to aid him in reducing his retail price by three cents per gallon, the six cent operating margin will be reduced by only approximately one cent per gallon. With price protection dealers are, therefore, much more willing to reduce prices so that they can increase or hold their volume.

The detailed procedure by which major oil companies grant price protection is illustrated by the price protection schedule shown in Table 7-1. It is based to a large extent on the plan used by Gulf Oil Company in one of its markets during 1970, and it is similar to plans used by Gulf and other majors throughout the country. The normal price (no support) for major-brand gasoline before taxes is shown to

TABLE 7-1

SAMPLE PRICE PROTECTION SCHEDULE[1]

Price Level	Retail Price (Before Tax)	Dealer Margin	Price Protection	Dealer Cost (Tank Wagon)	Jobber Margin	Jobber Assistance	Branded Wholesale
		Premium Grade Gasoline Dealer Price Schedule			Jobber Price Schedule		
Normal	29.9¢	7.5¢	0	22.4¢	4.50¢	0	17.90¢
Off 1¢	28.9	7.2	0.7¢	21.7	4.29	0.49¢	17.41
" 2	27.9	6.9	1.4	21.0	4.08	0.98	16.92
" 3	26.9	6.6	2.1	20.3	3.87	1.47	16.43
" 4	25.9	6.3	2.8	19.6	3.66	1.96	15.94
" 5	24.9	6.0	3.5	18.9	3.45	2.45	15.45
" 6	23.9	5.7	4.2	18.2	3.24	2.94	14.96
" 7	22.9	5.4	4.9	17.5	3.03	3.43	14.47
" 8	21.9	5.1	5.6	16.8	2.82	3.92	13.98
" 9	20.9	5.0	6.5	15.9	2.75	4.75	13.15
" 10	19.9	5.0	7.5	14.9	2.75	5.75	12.15
" 11	18.9	5.0	8.5	13.9	2.75	6.75	11.15
" 12	17.9	5.0	9.5	12.9	2.75	7.75	10.15
			Regular-grade Gasoline				
Normal	25.9¢	7.0¢	0	18.9¢	4.00¢	0	14.90¢
Off 1¢	24.9	6.7	0.7¢	18.2	3.79	0.49¢	14.41
" 2	23.9	6.4	1.4	17.5	3.58	0.98	13.92
" 3	22.9	6.1	2.1	16.8	3.37	1.47	13.43
" 4	21.9	5.8	2.8	16.1	3.16	1.96	12.94
" 5	20.9	5.5	3.5	15.4	2.95	2.45	12.45
" 6	19.9	5.2	4.2	14.7	2.75	2.95	11.95
" 7	18.9	4.9	4.9	14.0	2.75	3.65	11.25
" 8	17.9	4.6	5.6	13.3	2.75	4.35	10.55
" 9	16.9	4.5	6.5	12.4	2.75	5.25	9.65
" 10	15.9	4.5	7.5	11.4	2.75	6.25	8.65
" 11	14.9	4.5	8.5	10.4	2.75	7.25	7.65
" 12	13.9	4.5	9.5	9.4	2.75	8.25	6.65
			Subregular Gasoline				
Normal	24.9¢	6.5¢	0	18.4¢	3.75¢	0	14.65¢
Off 1¢	23.9	6.2	0.7¢	17.7	3.54	0.49¢	14.16
" 2	22.9	5.9	1.4	17.0	3.33	0.98	13.67
" 3	21.9	5.6	2.1	16.3	3.12	1.47	13.18
" 4	20.9	5.3	2.8	15.6	2.91	1.96	12.69
" 5	19.9	5.0	3.5	14.9	2.70	2.45	12.20
" 6	18.9	4.7	4.2	14.2	2.50	2.95	11.70
" 7	17.9	4.5	5.0	13.4	2.50	3.75	10.90
" 8	16.9	4.5	6.0	12.4	2.50	4.75	9.90
" 9	15.9	4.5	7.0	11.4	2.50	5.75	8.90
" 10	14.9	4.5	8.0	10.4	2.50	6.75	7.90
" 11	13.9	4.5	9.0	9.4	2.50	7.75	6.90
" 12	12.9	4.5	10.0	8.4	2.50	8.75	5.90

1. Basic data derived from Gulf Oil's price protection schedule for 1970 in a given market.

FIGURE 7-1

ILLUSTRATION OF WAY PRICE PROTECTION APPEARS TO AN OIL COMPANY

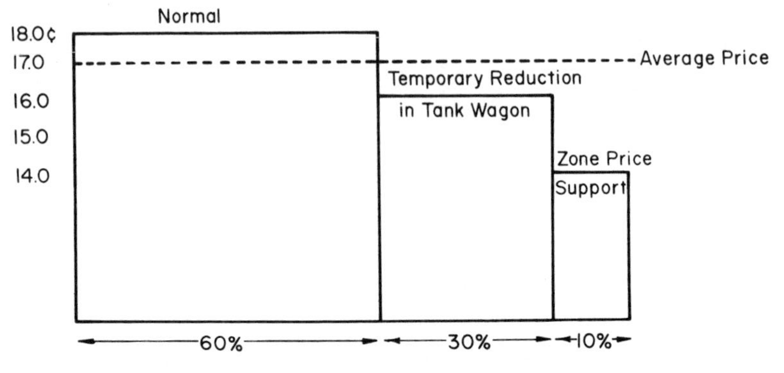

be 29.9¢ for premium, 25.9¢ for regular, and 24.9¢ for subregular with respective dealer margins of 7.5¢, 7.0¢, and 6.5¢ per gallon of gasoline. Regardless of the grade of gasoline, dealers receive 0.7¢ support for every 1.0¢ per gallon reduction in the suggested retail price of gasoline. In other words, the major oil company (or the jobber and major oil company if jobbers are used) absorbs 70 percent of the reduction in the retail price and the dealer absorbs 30 percent of the reduction, until the minimum guaranteed margin is reached of 5.0¢ on premium and 4.5¢ on regular and subregular. When the minimum margin is reached, the major oil company absorbs the full amount of any further decreases in suggested retail price.*

A broader view of the process by which major oil companies selectively use price protection to combat price competition is illustrated by the diagram in Figure 7-1. Assume that an oil company is operating

* One of the earlier price-protection plans was Texaco's Chicago Plan. It was a 20/80 percent plan while today most of the plans are 30/70 percent. The higher the sharing of the cost by the dealer, the greater the likelihood that dealers will not want to prolong price wars and will be anxious to have prices restored to normal.

throughout all of the fifty states. The normal dealer price is eighteen cents per gallon for regular gasoline. Sixty percent of the major's volume is being sold at eighteen cents per gallon, 30 percent is moving at an average of sixteen cents as a result of a temporary reduction in the tank-wagon price, and 10 percent is being sold at fourteen cents because of zone price protection. In those areas where major oil companies face active price competition, price protection is used to reduce dealer prices. Figure 7-1 illustrates another characteristic of the practice of granting price protection. While the normal price is eighteen cents per gallon, price protection draws the average price down to seventeen cents per gallon. Actually the level of support indicated in the example is apparently too low for 1970. According to a report of the Office of Emergency Preparedness one company submitting confidential figures had paid price support of 1.3¢ per gallon during 1970 and the *National Petroleum News* reported a larger oil company with price support that had averaged near 2.0¢ per gallon in 1970.[1] Price-protected markets and stations are being subsidized by the high tank-wagon price levels in the less competitive markets or by profits drawn from other levels of industry activity.

Price Protection Programs with Different Geographic Scope

Price protection is granted to dealers in two basic ways: (1) through temporary reduction in the dealer tank-wagon price (the price dealers pay for gasoline) or (2) through zone price support. Temporary reduction in dealer tank-wagon price normally involves giving the same level of price protection throughout a "broad trade area" such as, for example, Kansas City or even Detroit. In contrast, zone price support involves giving price protection to selected dealers in certain small areas within the broader trade area.

The two methods of granting price protection are illustrated by the recent pricing practice of the major oil companies in Detroit. From 1966 through the early part of 1969 the price that major-brand dealers paid for regular-grade gasoline was for the most part 16.5¢ per gallon with a suggested retail price of 35.9¢. Reportedly, during this period there was very little, or no, price protection given by the major oil companies. The price dealers paid for regular gasoline was then increased to 17.3¢ and 18.0¢ with the suggested retail price on regular gasoline increasing to 37.9¢ per gallon. Beginning at the end of June 1969, temporary reductions in the dealer tank-wagon price were generally made throughout most of Detroit, which caused a seesawing pattern of tank-wagon prices. Following the third increase in tank-wagon prices in Detroit in a little over a year, this time to 18.7¢, price-protection practices combined temporary reductions in the dealer tank-wagon prices and zone price supports. For example, the market leader, American Oil, established fifty different zones in Detroit and competitors were said to have as many or more.[2] Instead of prices being uniformally decreased throughout the city, there was great variance from one part of it to another. For example, the *Platt's Oilgram Price Service* (April 21, 1970) reported that regular-grade gasoline to consumers might range from a low of 29.9¢ with support to a normal of 38.9¢ without support.[3] This meant that some dealers were paying 12.4¢ per gallon of gasoline (tank-wagon price less price protection) while other dealers received no price protection and were paying 18.7¢.

The Detroit price wars, financed by price protection, were designed to curb price cutting by certain major-brand dealers and price marketers. The price wars were intended to make it unprofitable to sell gasoline by discounting prices. Two years of price wars have definitely diminished the number of stations discounting gasoline in Detroit. For example, many of the major-brand dealers that were selling on a price-discounted basis have been driven out of business.[4] A large major oil-company jobber who had many stations with cut prices found his supply contract cancelled and sold out. Also Tulsa, the largest private brander in the city, decided that it was in its best interest to sell out. Price protection was performing its function of "shaping-up the market" and convincing a large number of operators that it didn't pay to cut prices, and that the acceptable norm of competitive behavior was in terms of the nonprice marketing of gasoline.

Spread in Use of Price Protection

The practice of granting price protection became increasingly common in the later 1950's and fed the widespread price wars of the early 1960's. The growth of price protection during this period is illustrated by the experience of Apco, a large integrated mid-continent independent.

> In the case of our company or its predecessor, price protection was practically unheard of as late as 1957 but it grew in each succeeding year. Here follows a tabulation of our cost of price protection for the last 6 years:
>
> | | | | |
> |---|---|---|---|
> | 1959 | $ 877,000 | 1962 | $2,843,000 |
> | 1960 | 1,339,000 | 1963 | 2,641,000 |
> | 1961 | 1,795,000 | 1964 | 4,774,000 |
>
> In March of this year we reached an all time high. In the single month price protection cost us $675,000. That was the amount of money that was necessary for us to rebate back to our dealers and distributors to compensate for the low market quotations.
>
> Chairman DIXON. Off posted tank-wagon prices?
>
> Mr. RODMAN. Off our publication price that we sell.
>
> As a trend against earnings it is interesting to note that in 1959 the price protection granted by our company equaled less than one-sixth of the company's income before tax, but that by 1964 the price protection equaled more than three times the company's income before tax. We market approximately 300 million gallons of gasoline per year; hence, the above figures show that during 1964 price protection cost us in excess of 1½ cents per gallon.[5]

The destructive gasoline price wars of the first half of the 1960's came to an abrupt halt in many markets on March 17, 1965. The catalyst in ending the price war was Texaco, and the way it was done was relatively simple. Operating in all fifty states, Texaco announced that it was making a national price restoration. In other words, Texaco was withdrawing its price protection around the country. Texaco's move was rapidly supported by Humble, American, Mobil, and most of the other major oil companies. One of the principal reasons Texaco and Esso, the two giants of the industry, had for making this decision was that depressed market prices were beginning to adversely affect their earnings.

The withdrawal of price protection by the major oil companies meant that prices in most of the badly depressed markets would return to normal levels. For a time the practice of widespread cross-market subsidization and vertical subsidization would cease. The dispute over the one-cent differential was dropped, and price differentials between the majors and price marketers rose from two to five cents per gallon for regular-grade gasoline.

With the March 17, 1965, price restoration many troubled markets of the earlier 1960's greatly improved. For three or four years prices generally remained strong. However, in the latter 1960's price protection began once again to be used with increasing frequency in many of the previously hard-hit markets.

During 1969 and the first half of 1970 price-war conditions returned to many of the large mid-continental and midwestern cities. In Detroit, Chicago, Milwaukee, Kansas City, Denver, Houston, Dallas, and San Antonio price spirals reappeared and markets again became chaotic as the majors granted more price protection to their dealers. Consumer prices once again dropped eight, ten, or even fourteen cents per gallon over a period of a few weeks. The wide fluctuation in prices in these and other markets is illustrated by the excerpts from the industries' leading price authority the *Platt's Oilgram Price Service* which follows.

FEBRUARY 9, 1970

There may have been a period in the past four or five years when retail gasoline prices were in a generally worse shape than they're in right now. But it would take a sharp memory to recall it. Things look that bad to midcountry marketers at the moment.

Hardly a major metropolitan area is benefitting from full established dtw [dealer tank wagon] postings today, for some, such as Detroit and Kansas City and Denver, subnormalcy has been virtually a way of life for the past few years.

But the current price crisis is relatively recent to some other large markets, such as St. Louis, which had been normal for five years until this past week's 4¢–5¢ gal retail break.

Partial lists of larger markets where general dtw allowances currently are supporting subnormal pump prices includes:

Milwaukee, retail off 7¢ to 10¢ gal; Minneapolis-St. Paul 2¢; Detroit 8¢; St. Louis 4¢–5¢; Kansas City 6¢; parts of Chicago 2¢–4¢; Peoria, Ill. 5¢; Danville, Ill. 9¢; Springfield, Ill. 5¢; Galesburg, Ill. 11¢; Chrisman, Ill. 14¢; Indianapolis 1¢; Evansville, Ind. 7¢; South Bend, Ind. 3¢.

Also, Des Moines, Iowa 8¢ to 10¢; Rapid City, Sioux Falls and Watertown, S.D., all off 10¢; Fargo, N.D. 10¢; Grand Forks, N.D.

4¢; Atchison, Kan. 10¢; Hutchinson, Kan. 9¢; Denver 8¢ to 11¢; Pueblo, Colo. 9¢; Colorado Springs, Colo. 7¢.

Also Tucson, Ariz. 8¢; Phoenix 1¢; Salt Lake City 6¢; Provo, Utah 7¢; Cheyenne, Wyo. 8¢; Butte, Mont. 2¢; Billings, Mont. 2¢; Oklahoma City 7¢ to 9¢; Ft. Worth, Tex. 1¢ to 10¢; Houston 1¢ to 8¢.

JUNE 18, 1970
MID-CONTINENT/MIDWEST

As the worst war-torn sector of the U.S., midcountry will be most vitally affected by Phillips' sweeping gasoline price restoration of June 22 (story this issue). Most of the largest-volume markets in the interior currently are in throes of serious retail gasoline wars, e.g., Detroit; Chicago; Milwaukee; Minneapolis-St. Paul; Indianapolis; Kansas City; Lincoln, Neb.; Houston; Tucson, as well as scores of medium-size markets.[6]

Zone Price Support

Zone price support often supplements wide-area price protection. It is a more refined application of the general practice of granting price protection. Instead of reducing the price that dealers pay for gasoline throughout wide areas (e.g., Washington, D.C.), zone pricing permits lower prices to be programmed into particular parts of cities.

From trade reports and discussions with gasoline marketers it appears that Shell Oil Company has been a leader in developing the practice of zone pricing. Shell introduced its TAP (Trade Area Plan) plan in Chicago and Detroit early in 1961. The plan, as explained, was designed to "stop the volume drain" of the price marketers and to "narrow the price differential" between the price marketers and the majors from more than two cents to two cents or less.[7] What Shell did was to divide its markets into a large number of small zones. Shell then was in a position to clamp down on strong price competitors in certain areas while leaving its price structure intact in less competitive areas.

Zone pricing then became a means by which majors could regulate, to a degree, the growth of price marketers in areas where they had made, or were making, significant inroads. The control process centered on the price marketers' gross profit. Shell's TAP plan, Texaco's Chicago plan, and American's TACA (Trade Area Competitive Allowance) plan all affected the price marketers' volume and margin. By narrowing the price differential between the price marketers and major brands from three or more cents to two cents or less per gallon, the majors could drain off the price marketers' volume. The example in Table 7-2 shows the differenital has been reduced from three to two

TABLE 7-2

Process by Which Zone Pricing Plans Are Used to Control the Growth of the Price Marketers

		After "TAP"		
	Before "TAP" 3¢ Differential	Reduced Differential to 2¢ or Less	Reduced Margin	Reduced Margin and Differential
Monthly gallonage	100,000	75,000	100,000	75,000
Gasoline margin	8¢	8¢	6¢	6¢
Gross profit	$8,000	$6,000	$6,000	$4,500
Reduction in gross profit		$2,000	$2,000	$3,500

cents or less per gallon resulting in volume falling off by 25 percent and gross profits reduced by $2,000. If the price marketer chooses to reduce his price to maintain the differential and to hold his volume he can do so by narrowing his margin. In the example he cut his margin from eight cents to six cents per gallon, held his volume, and reduced his gross profit by $2,000. In reality both of these effects normally work together when a major uses zone pricing against a price marketer. The price marketer loses his volume as well as his margin. The final example shows both volume and gasoline margin decreasing by 25 percent which results in gross profit declining from $8,000 to $4,500, a reduction of $3,500. Furthermore, the example also shows what frequently happens when prices fall by eight cents per gallon—no operating margin remains to pay employees and cover other operating costs.

Examples of Zone Pricing

Two markets that illustrate the practice of using zone pricing to contain price marketers are Washington, D.C. and San Francisco (including the West Bay area through San Jose). In Washington, D.C. price marketers account for only 7 percent of the stations (54 out of 694 stations) surveyed on the major arteries. The intensive market coverage of the majors puts them in a position to lower prices in certain of their stations near the price marketers to reduce the price marketers' ability to compete. In those areas where the price marketers are not represented, or are unimportant in terms of volume, the majors generally maintain higher price levels. While the majors could not generally afford to lower prices throughout the District of Columbia, they could develop a satisfactory overall system-level price by averaging their subsidized stations with their higher priced stations. The way in which the majors zoned the price marketers in Washington, D.C. was generally to (1) reduce the price at selected major-brand stations adjacent to or near strong price marketers by one to three cents per gallon below the normal major-brand price throughout the market and (2) post the "specially" low price of major-brand gasoline in the competitive pockets. (Data illustrating the practice of zone pricing in Washington, D.C. are presented in Appendix C.)

The price studies carried out in the San Francisco, West Bay, and Greater San Jose area (see Table 7-3) reveal many examples of deep-cut zone pricing. In those zones where strong and aggressive price marketers were located major-brand prices would frequently be 3.0¢ to 5.0¢ lower than in nearby areas where price marketers were not located or were unimportant. For the first two groups of stations (numbering thirty-five and twenty-seven respectively), located in the southern part of San Francisco or just south of San Francisco, prices were high and stable and there were no strong price marketers in this area. The prevailing major-brand price was 36.9¢ per gallon for regular with the next largest group of stations selling regular for 37.9¢. As we continue south on the El Camino Real road we find sixteen stations in the Palo Alto area—thirteen majors and three price marketers. Several of the majors had cut their prices by 3.0 to 5.0¢ per gallon below those shown in the previous areas to compete with some aggressive low-price independents. As we proceed south along the El Camino Real road, we find in the Mt. View area of the West Bay eighteen major-brand stations and no price marketers and the median major-brand price has increased by two cents from the previous area. Note also Shell's prices in this zone—one of its stations was selling regular for 34.9¢ per gallon and three were selling it for 36.9¢ per gallon with none posting their prices. In the previous zone, where there were strong price marketers, Shell stations were selling regular gasoline for 32.9¢, 33.9¢, and 34.9¢ per gallon, and of these stations two were posting their price. Moving further south along the El Camino Real road into Sunnyvale, we find

TABLE 7-3

Cluster Analysis in San Francisco and Communities South of the City

Regular	28.9¢	29.9¢	30.9¢	31.9¢	32.9¢	33.9¢	34.9¢	35.9¢	36.9¢	37.9¢
Premium	31.9–32.9¢	32.9–33.9¢	33.9–34.9¢	34.9–35.9¢	35.9–36.9¢	36.9–37.9¢	37.9–38.9¢	38.9–39.9¢	39.9–40.9¢	40.9–41.9¢
Differential	3–4¢	3–4¢	3–4¢	3–4¢	3–4¢	3–4¢	3–4¢	3–4¢	3–4¢	3–4¢
Mission Junipero Alemany Skyline 35 stations (0–35)							GUL 1	SH	MOB CHE	UN (+2¢)
Mission Junipero Alemany Skyline 35 stations (0–35)							PHI PHI	MOB SH TEX SH CH	RIC RIC MOB PHI PHI RIC RIC GUL2 CHE CHE PHI	CHE CHE PHI SH UN CHE CHE RIC PHI CHE UN CHE
El Camino Real Road in Milbrae and Burlingame 27 stations (3–24)								CHE	MOB CHE	
El Camino Real Road in Milbrae and Burlingame 27 stations (3–24)				HANCOCK	REGAL	ROCKET	TEX	CHE CHE	MOB ESS UN RIC CHE PHI SH PHI UN CHE UN RIC PHI TEX	PHI SH UN GUL1
El Camino Real Road in Pal Alto 16 stations (3–13)		HUDSON PB			ESS SH	PHI SH GUL2	RIC CHE			PHI
El Camino Real Road in Pal Alto 16 stations (3–13)			BEACON			MOB	SH		MOB CHE	CHE (+1)
El Camino in Mt. View 18 stations (0–18)				TEX	ESS	TEX	GUL1 SH PHI	CHE UN CHE	SH SH CHE RIC CHE	
El Camino in Mt. View 18 stations (0–18)			ESS					UN	SH CHE	
El Camino Real Road in Sunnyvale 18 stations (3–15)				DISC. STORE	SH AM	RIC TEX GUL3	GUL2		CHE	RIC (+2)
El Camino Real Road in Sunnyvale 18 stations (3–15)			DOUGLAS (+2)	HANCOCK PHI	SH	TEX	MOB	TEX	CHE UN	

TABLE 7-3 (continued)

Regular	28.9¢	29.9¢	30.9¢	31.9¢	32.9¢	33.9¢	34.9¢	35.9¢	36.9¢	37.9¢
Premium	31.9–32.9¢	32.9–33.9¢	33.9–34.9¢	34.9–35.9¢	35.9–36.9¢	36.9–37.9¢	37.9–38.9¢	38.9–39.9¢	39.9–40.9¢	40.9–41.9¢
Differential	3–4¢	3–4¢	3–4¢	3–4¢	3–4¢	3–4¢	3–4¢	3–4¢	3–4¢	3–4¢
San Antonio Rd. and Foothill Expy. 27 stations (0–27) (No Stamps)							UN	CHE SH MOB RIC UN CHE RIC	SH MOB MOB UN GUL2	CHE CHE SH
(Stamps)				TEX			RIC ESS RIC	CHE TEX CHE	CHE MOB UN	MOB

Codes: No Stamps (above dashed line) / Stamps (below dashed line)

Major brands: CHE—Chevron; SH—Shell; ESS—Esso; PHI—Phillips; TEX—Texaco; UN—Union; RIC—Richfield; MOB—Mobil; GUL—Gulf

Secondary major brands: ROCKET—Richfield; DOUGLAS—Conoco

Integrated independent brands: REGAL—Signal Oil & Gas; HANCOCK

Price marketers: PB—private brand (general); HUDSON; BEACON; DISCOUNT STORE

GUL 2—2 is tane price less than regular.

—— —indicates street posting of price.

( ) —indicates premium differential is other than indicated by category headings.

a group of eighteen stations including three volume price marketers (Whitefront Discount Store, Conoco's price-brand, Douglas, and a Hancock station of the Signal Oil Company). Several of the majors here have reduced their prices 3.0¢ or 4.0¢ below those in areas where there was limited or no price competition. Off the El Camino Real road where there were no disturbing price marketers, prices were at their normal high level. This is illustrated by the pricing of stations located on the San Antonio road and off the Foothill Expressway. Only one of these twenty-one stations was posting a price and the median price in the area had shifted upward to 35.9¢ per gallon for regular and 40.9¢ per gallon for premium with several stations 1.0¢ or 2.0¢ higher.

ZONE BOUNDARIES AND THE PRACTICE OF FEATHERING

ANOTHER ASPECT of zone pricing which helps one to understand the practice is to consider how the zone boundaries are established. Are the zone boundary lines very broad so that they encompass natural trade areas? Or, on the other hand, are zones drawn very tightly, encompassing only a few stations that are arbitrarily placed in particular groups? From observing the market places where zone pricing has been employed, from studying the zone network of a major oil company, and from information obtained about zone practices of other majors, it is reasonably clear that zone boundaries are arbitrarily drawn and do not reflect natural trade areas. For example, many of the arteries leading out of Washington, D.C. and the El Camino Real road running south of San Francisco carry huge amounts of commuter traffic several miles daily in each direction. Yet, these arteries are cut by several artificially established zones.

Cities like Chicago are frequently divided into hundreds of zones which are used to regulate price competition. The artificial way in which zones are established can be observed from studying an American Oil Company's TACA map for the northern section of Chicago. The area shown generally involves a distance of fifty blocks north to south and forty blocks east to west (see Figure 7-2). Instead of the area being designated as one natural trade area bounded on the East by Lake Michigan and on the West by the Kennedy Expressway, it is divided into twenty-three arbitrarily drawn zones. A majority of the commuter traffic probably passes on a daily basis through two or more

FIGURE 7-2

Zone Pricing Map for American Oil in a Northern Section of Chicago—1969

THE SMALL CIRCLED NUMBERS REPRESENT STANDARD OIL OF INDIANA STATIONS.

of these zones. For example, on Western Avenue (2400 West), a major north-south artery, from Montrose (4400 North) to Peterson (6000 North), a distance of sixteen blocks, a driver passes through four zones (see dotted line designated as A). Or a driver starting from Lincoln and Peterson and going to the expressway at Hollywood, which many drivers do on a daily basis, passes through five zones (see dotted line designated as B).

Besides the deleterious effect that zone pricing has on price marketers, it also puts the major oil companies in a position of discriminating against certain of their own dealers. Since zones are not generally developed around natural marketing areas, the practice of granting price protection results in certain nearby operators of the same brand receiving different prices. The practice of feathering has been suggested as a technique that the major oil companies might employ to reduce the impact of the price discrimination between their dealers resulting from zone pricing. The idea of feathering is to graduate outward from a low-price zone the amount of the price subsidy given, so that dealers in adjacent trade areas will not have materially different costs, which will impair their ability to compete. However, the gasoline-price surveys carried out in Washington, D.C. and in San Francisco showed that frequently there was little graduation in price differences among major-brand dealers of the same brand located a short distance apart. Similarly an analysis of American Oil Company's zone pricing for a northern part of Chicago showed rather significant differences in price support between adjacent and nearby zones. For example, consider the difference in the amount and magnitude of the zone price support for zones 322, 323, and 324 on Foster Street (see

TABLE 7-4

COMPARATIVE NUMBER OF DAYS OF TACA PRICE SUPPORT GRANTED TO ZONES 322, 323, AND 324

TACA Zone Number	1.3¢ Support Suggested Retail Price Reduction of 2.0¢	2.0¢ Support Suggested Retail Price Reduction of 3.0¢	2.7¢ Support Suggested Retail Price Reduction of 4.0¢	Total Days of Support
322	50	0	0	50
323	27	107	0	134
324	77	97	39	213

Table 7-4 and Figure 7-2). Certain stations in adjacent or nearby zones were given zone price protection permitting them to reduce prices two to four cents per gallon below other stations in the general area.

METER READING PROGRAMS

PRICE PROTECTION granted during the first half of the 1960's ran into hundreds of millions of dollars. With such a huge amount of money being paid out in price subsidies, schemes were developed to beat the system. In some cases major-brand jobbers, jobbers of the small integrated companies, and large-volume dealers owning and operating their own stations found ways to turn price protection to their own advantage. Some jobbers turned in falsified claims for price protection to oil companies for assistance allegedly granted certain dealers. For example, a jobber might claim support for 100,000 gallons of gasoline to ten stations at two cents per gallon, while only half that amount of subsidy was actually given. The jobber would gain $1,000 which could be used to increase his profit or to cut his prices and become more competitive. Another abuse of price protection was cross hauling gasoline from price subsidized markets to high price markets. For example, a major oil company discovered that one of its jobbers was buying price-supported gasoline in the low-priced Buffalo market and transporting it 300 miles to the high-priced Connecticut market via the interstates.[8] Similarly a rural jobber for a major oil company located in a chronically depressed area in Missouri was found transporting price-supported gasoline into St. Louis and upsetting the market. Another cross-hauling practice involved dual brand jobbers—those operating both branded and unbranded stations. When the price-supported major-brand gasoline fell below the unbranded price for gasoline, some of the price-subsidized branded gasoline moved through the unbranded outlets; the reverse practice was followed when the unbranded market price was considerably below the branded wholesale price. A similar practice has been for branded jobbers to sell price-subsidized gasoline to unbranded operators when conditions favored such transactions.

Another abuse of price protection has been that some dealers with large storage capacities (e.g., 80,000 to 100,000 gallons) have been known to play the market. When price subsidies result in low-price gasoline, they keep their storage tanks filled. They then have an extra long operating margin when prices return to normal which they can either pocket or use to cut price to build volume. The practice is illustrated by a customer of a Fina jobber who opened a new station with a storage capacity of 80,000 gallons. According to the supplying jobber, the large storage capacity was "the best investment the dealer could make and he'll pay for those tanks in six months as a result of price protection."

In an attempt to stop some of these abuses, several of the major oil companies in mid-1963 implemented meter-reading programs. A representative of the oil company, or its jobber (if used), was supposedly required to read the gasoline meters of dealers authorized to get price protection at the time of price changes and to certify that each dealer was entitled to a certain amount of subsidy. In general the meter-reading programs gave the major oil companies the control they needed to direct price protection in the areas and localities for which the price subsidies were intended.

SELLING BELOW REASONABLE COSTS

PRICE-PROTECTION programs are designed so oil companies can sell low-price gasoline in certain areas to combat price-discounted competition. The major oil companies cannot generally afford to sell at the average price they receive in markets where they are giving price protection. However, when prices from less competitive markets are averaged with the prices from subsidized markets, the major oil companies generally realize a satisfactory overall price for their products.

In many price-war markets, price protection results in major oil companies selling gasoline below reasonable cost. The price wars in Detroit of the latter 1960's and early 1970's provided many examples of this. An approximation of major oil company refinery netback on the sale of regular gasoline at different retail prices is presented in Table 7-5. Refinery netback (i) is calculated in the following manner:

Suggested retail price
Less: Taxes
Dealer margin
Jobber margin
Distribution cost
Branding cost
Refinery Netback

A comparison is made of refinery netback and the Gulf Coast Low Price for regular-grade gasoline. The Gulf Coast Low Price is frequently the price at which volume purchasers exchange gasoline, and is considered to be the low price at which gasoline is normally traded.[9]

The normal, no-subsidy, major-brand price in Detroit during the second quarter of 1970 was 38.9¢ per gallon. However, prices were normal for only short periods of time and were more often 5.0¢ to 10.0¢ (at 34.9¢ to 28.9¢), or more, below normal. At these low prices refinery netbacks were 0.5¢ to 4.5¢ below the Gulf Coast Low—the benchmark used to indicate when prices have fallen below reasonable

TABLE 7-5

Approximation of Refinery Netback for Major Oil Companies for Regular Gasoline at Different Retail Prices in Detroit, Michigan, 2nd Quarter, 1970

Suggested Price for Regular Gasoline (a)	Sales Tax and State and Federal Fuel Taxes[1] (b)	Dealer Margin[2] (c)	Jobber Margin[3] (d)	Est. Delivery Terminal and Pipeline Costs[4] (e)	Estimated Branding Costs[5] (f)	Refinery Netback (g)	Gulf Coast Low Price Regular Gasoline[6] (h)	Refinery Netback Less Gulf Low Price (i)	Price Protection Given by Oil Company[7] (j)
						[a − (b,c,d,e,f)]		(g − h)	
38.9	12.26	7.94	3.75	2.00	1.00	11.95	10.5	1.45	—
37.9	12.23	7.67	3.54	2.00	1.00	11.46	10.5	0.96	0.49
36.9	12.19	7.41	3.33	2.00	1.00	10.97	10.5	0.47	0.98
35.9	12.15	7.15	3.12	2.00	1.00	10.48	10.5	−0.02	1.47
34.9	12.11	6.89	2.91	2.00	1.00	9.99	10.5	−0.51	1.96
33.9	12.07	6.63	2.75	2.00	1.00	9.45	10.5	−1.05	2.50
32.9	12.04	6.36	2.75	2.00	1.00	8.75	10.5	−1.75	3.20
31.9	12.00	6.10	2.75	2.00	1.00	8.05	10.5	−2.45	3.90
30.9	11.96	5.84	2.75	2.00	1.00	7.35	10.5	−3.15	4.60
29.9	11.92	5.58	2.75	2.00	1.00	6.65	10.5	−3.85	5.30
28.9	11.88	5.32	2.75	2.00	1.00	5.95	10.5	−4.55	6.00
27.9	11.84	5.06	2.75	2.00	1.00	5.25	10.5	−5.25	6.70
26.9	11.80	5.00	2.75	2.00	1.00	4.35	10.5	−6.15	7.60
25.9	11.77	5.00	2.75	2.00	1.00	3.38	10.5	−7.12	8.57
24.9	11.73	5.00	2.75	2.00	1.00	2.42	10.5	−8.08	9.53

1. Sales tax is 4%, state fuel tax is 0.7¢ and federal is 0.4¢ per gallon.
2. Dealer margin reduces 0.3¢ for every one-cent reduction in retail, less a sales tax adjustment of 0.04¢. Thus, the dealer loses approximately 0.26¢ for every one-cent reduction in suggested retail price. Once the dealer margin falls to 0.5¢, it no longer is reduced; the "stop-out" has been reached.
3. The jobber's margin declines 0.21¢ for every one-cent reduction in the retail price. The jobber's stop-out margin is 2.75¢.
4. Local delivery is calculated at 0.65¢ per gallon, pipeline tariff from Beaumont, Texas, to Detroit is 1.48¢ and the terminal charge is 0.3¢. Thus the total is 2.43¢. However, 0.2¢ is used in the approximation as a low cost estimate for this activity.
5. A low estimate of the branding costs that were described in Chapter 2 and include advertising, credit card cost, and sales promotion program.
6. *Platt's Oilgram Price Service.*
7. When an oil company operates through a jobber, the oil company gives price protection of 0.49¢ for every one-cent reduction in the suggested retail price. The jobber gives up 0.21¢ and the dealer 0.3¢ for every one-cent reduction in the suggested retail price. Once the jobber's stop-out is reached (33.9¢) the oil company's price protection increases to 0.7¢ for every one-cent reduction in the suggested retail price. When the dealer's stop-out is reached (27.9¢), the oil company's price protection increases to one cent per gallon.

cost. When retail prices are 5.0¢ to 10.0¢ below normal, the major oil companies using jobbers are giving price protection of 2.0¢ to 6.0¢ per gallon of gasoline (j). Frequently, at these low prices, refinery netback will not cover processing costs, and at times it will be below the cost of crude oil to the refineries.

SUMMARY

THE PRICE WARS described in Chapters 5 and 6 have been financed by the various price-protection plans of the major oil companies. These price protection programs are used to defend the majors' nonprice methods of marketing gasoline against price competition. The majors selectively reduce prices in those areas where they face strong price marketers. By using price protection in this manner the majors are able to discipline and regulate competitors selling gasoline on a price-discounted basis. Often the granting of price protection results in deep price cuts that are below the reasonable costs of doing business. In this respect price protection is a massive subsidization plan with revenues from less competitive markets, and other levels of business activity, being used to finance sales below reasonable costs in highly competitive markets.

Were it not for price protection, intertype competition would force many changes in gasoline marketing that would improve efficiency and lower the price of gasoline. Major oil company executives recognize that gasoline marketing would be more efficient without price protection. Several of the representatives of major oil companies (American, Mobil, and Sun) stated during the 1965 FTC hearing on gasoline marketing that if price protection were eliminated large numbers of service stations would be forced to close. In 1970, Tom Sigler, vice president of marketing, Continental Oil Company, pointed out much the same relationship:

> Sigler told distributors that the system helps "perpetuate the inefficient and obsolete outlets with the incredible philosophy that—if one has to go broke, then we will all go broke."
>
> The Conoco vice president said the faulty system is under-mining "the long-term health and profitability of the entire marketing segment of our business . . . economic health will come to our branch

of the industry when we face up to this situation and refuse to use artificial respiration to breathe new life into dying outlets." [10]

It seems rather obvious that if price protection schemes were eliminated, large numbers of unneeded stations would close and supply and demand would approach a more balanced condition. This would mean that marketing costs would fall and that the level of services would likely improve as economies of scale of the larger stations started to take hold. It also means that the long-run health and profitability of the entire marketing phase of the business would improve. On their own, however, it is doubtful that the major oil companies will stop the questionable practice of granting price subsidies. As long as one significant oil company grants price protection, its competitors will be compelled defensively to continue the practice.* A second reason that the major oil companies will not cease the practice of price protection on their own is that they are fearful that price marketers would grow at their expense. This would very likely be the situation in the short run, and it is in the short run (one or two years) that most managements operate. However, in the long run the elimination of price protection and the reduction of major marketing costs could check the growth of price marketers and eliminate the need for price-subsidization programs to regulate artificially price competition.

* Both Conoco and Phillips experimented with dropping their price protection programs in the latter part of 1970, but were forced to return to the practice when competitors continued to give price protection.

8 / *Buy-outs of Price Marketers*

ONE OF THE MOST STRENUOUS TYPES OF COMPETITION that exists in gasoline marketing is intertype competition (see Chapter 4, pp. 107–08). As a result of intertype competition, drivers in many markets have a choice of distinctly different retail operations from which they can purchase gasoline. Furthermore, intertype competition is also very important since it often represents a challenge to the older, dominant method of doing business, and as a result has the potential of forcing change in staid marketing practices.

Selective price wars, financed by price protection, as discussed in Chapters 5 to 7, has been one of the ways the major oil companies have thwarted and otherwise checked intertype competition. Another method employed to reduce intertype competition has been the buy-outs of price marketers. In many cases, price wars, price protection, and the buy-out have been used as complementary techniques. Price wars have frequently resulted in the "softening-up" of price marketers and have convinced many operators that it would be in their long-run best interest to sell out.

To assess the impact of the buy-outs of price marketers on intertype competition, it is useful to divide the mergers involving price marketers into three principal categories:

1. buy-outs of price marketers with conversion to major brands;
2. buy-outs of price marketers with operation as major secondary brands; and
3. buy-outs of price marketers by integrated independents.

Purchase of Price Marketers with Conversion to Major Brands

There has probably been no more direct approach to eliminating price marketers and reducing intertype competition than by buying them out and converting them to the major-brand style of operation. One of the leaders in the buy-out of price marketers has been Humble Oil and Refining Division of Standard Oil of New Jersey, the world's largest oil company. As Esso expanded into the Midwest and Far West it also purchased large numbers of significant price marketers, ranging in size from single-station operators to chains with more than 200 stations. Eventually the private brander's identification was pulled down and the Enco brand and way of operating substituted in its place. A list of some of the independents purchased by Humble Oil and Refining is shown in Table 8-1.

TABLE 8-1

A Partial List of Private Branders Purchased by the Humble Division of Standard Oil of New Jersey

Name of Acquired Company	Date	Brand Sold Prior to Acquisition	Market Area	Number Stations
Oklahoma Oil Co.	1956	Oklahoma	Chicago, Ill.	76
Perfect Power Co.	1956	Perfect Power	Chicago, Ill.	47
Gasteria, Inc.	1958	Gasteria Hoosier Pete Bonded	Ind., Ill., Iowa, and Ky.	263
Five Star Oil Co.	1959	Five Star	Phoenix, Ariz.	Unknown
Thompson Petroleum Products, Inc.	1959	Wisco	Racine, Wisc. Rockford, Ill.	Unknown
Major Gas Stations, Inc.	1960	Wilshire Major	Nevada and Arizona	Unknown
Sun Flash Oil Co.	1961	Sunflash	Columbus, Ohio	Unknown
Watson Oil Co.	1961	Hancock	Miami, Fla.	Unknown
Regal Petroleum Corp.	1961	Regal	Eastern, Fla.	Unknown
City Oil Co.	1961	O.K.	San Francisco, Calif.	Unknown
Economic Services, Inc.	1961	Walker	Los Angeles, Calif.	Unknown
The Petroleum Products Co.	1961	Red D	Stockton, Calif.	Unknown
Senco, Inc.	1962	Senco	Atlanta, Ga.	Unknown
Southern Oil Co.	1962	Deems	Florida	Unknown
Magnum Oil Co.	1962	Magnum	Los Angeles, Calif.	Unknown
Orbit Stations, Inc.	1962	Orbit	San Bernadino, Calif. Riverside, Calif.	Unknown

Source: In part from FTC study of acquisitions of price marketers by major oil companies.

The Humble mergers were presumably allowed on the ground that they were market-expansion mergers. Supposedly these mergers would not diminish competition since Jersey was not a significant competitor in most of the markets where it purchased private branders. Overlooked was the damage these mergers did to intertype competition. In most cases Jersey was not buying a similar type of competitor and merging the two operations. Instead it was purchasing an arch rival of the major method of doing business and converting it to the majors' style of operation. In essence what many of these mergers and conversions amounted to was the elimination of high-volume competitors selling price-discounted gasoline. That is, the mergers represented a reduction of intertype competition.

For example, consider Humble's purchase and conversion of Oklahoma and Perfect Power with 123 stations combined centered in Chicago and its purchase of Gasteria's 263-station, four-state operation with headquarters in Indianapolis. The Oklahoma and Perfect Power stations were large, high-volume price marketers. They normally priced one-half to one and one-half cents below the majors and increased the differential still more with trading stamps (which the majors were not using) and other promotions such as free Sunday newspapers and a variety of give-aways. Often "Gas for Less" signs were prominently displayed at the stations.[1] The various Gasteria brands similarly sold for one to two cents less than the majors and frequently gave cash redemption coupons worth approximately another penny.[2]

Following the purchase of Oklahoma, Perfect Power, and Gasteria, Humble gradually changed their pricing policy to sell with other majors,[3] and then converted the brands to Enco with its primary nonprice method of doing business. In the process Standard Oil of New Jersey eliminated three of the largest marketing chains in a five-state area and reduced the vitality of intertype competition substantially. Oklahoma and Perfect Power did close to 10 percent of the gasoline business in Chicago and accounted for approximately one-third of all independent volume.

A Lessening of Intertype Competition

Such mergers and conversions of price marketers in cities like Chicago have definitely reduced the average driver's opportunity to purchase gasoline at discounted prices. There are several reasons why such mergers often permanently damage intertype competition. One of the reasons is the limited availability of good locations. Supposedly this was in part Humble's rationale for buying private branders—to get locations that would otherwise be difficult or impossible to acquire. As a result in many parts of large cities where price marketers have been purchased they may never return. Another reason that the buy-out and conversion of significant price marketers does permanent damage to intertype competition is associated with the slow business development process itself. Many of the price-discount chains that Standard Oil of New Jersey purchased and converted had developed over long periods of time (e.g., fifteen to thirty years). The significant size to which some of the price marketers had grown put them in a position to operate effectively in the big cities. Real estate development alone in the major cities requires sizeable operations that can afford to pay high real estate prices. Business is an accumulative process and those already moving have the momentum to go ahead. Frequently the going concern has enough stable business to allow it to make the longer-run investments which new entrants could not undertake. Furthermore, the industry structure (see Chapter 9) places major obstacles in the way of potential new entrants who might otherwise be tempted to replace price marketers eliminated by mergers.

Intertype competition is also damaged because one buy-out is frequently followed by another. Following Jersey's move Gulf officially purchased in 1968 the large Bulko chain with approximately 150 stations operating in Chicago and throughout Illinois, Indiana, and Wisconsin. Plate 8-1 illustrates the process of phasing out the Bulko brand and replacing it with Gulf. As a result of purchases of price marketers by Jersey, Gulf, and others, and their conversion to major brands and to the majors' methods of doing business, the driver does not really have much choice but to buy major-brand gasoline in most parts of Chicago. What has happened in Chicago has happened in other markets.

While Jersey was making its big penetration in the upper Midwest by purchasing price marketers, Gulf was doing much the same thing on the West Coast. Wilshire Oil Company of California in which Gulf had a sizeable investment, purchased large numbers of price marketers in the latter 1950's and early 1960's. In 1955 Wilshire Oil Company, under the direction of Robert O. Anderson (now chairman of the board of Atlantic-Richfield) started putting together a conglomerate of West Coast price marketers. Notable among the purchases of price marketers were the 172 Sunset International stations and the 21 stations of the Armour Oil Company. When Gulf officially took over Wilshire in 1960, Wilshire had 600 outlets in Arizona, California, and Nevada and a 33,000 barrel-a-day refinery in the Los Angeles basin.[4] In 1965 Wilshire with 971 stations (more acquired by mergers) disappeared from the scene as the stations were converted to Gulf. The West Coast conglomerate of independents ceased to exist.[5] The West Coast independents were gobbled up not only by Gulf but by other majors that sought to extend their position by buying out price marketers. By the time the buy-outs were completed in the early 1960's, an article in the *National Petroleum News* commented that, "If you could buy all the independents left, . . . you couldn't make a dent." [6]

MAJOR SECONDARY BRANDS

SEVERAL OF THE MAJOR OIL COMPANIES acquired or otherwise developed secondary brands in the latter 1950's and early 1960's. Table 8-2 shows that several significant price marketers were purchased by the major oil companies and operated for a period as major secondary brands. Continental Oil Company alone purchased more than 700 stations and along with Standard Oil of New Jersey and Gulf has been a major acquirer of price marketers.

The table also reveals one of the concerns about major secondary brands. Are the secondary brands simply a temporary phase in the process of converting a price marketer to a major brand, or are they long-run business propositions which will continue to be operated using the strong price proposition that existed before the merger? In Gulf's case the secondary brand was only a temporary phase as was true with the three other primary purchases of price marketers by Gulf. The same was true for Continental's Douglas, Direct, and Mileage brands which have been replaced by Texaco and Conoco trademarks and the major method of operation. As a result of the conversion of the Douglas brand to Texaco in 1971, the 1965 change of Wilshire to Gulf, and the withdrawal of Signal Oil and Gas from marketing in 1971 (brands Hancock, Regal, etc.) intertype competition has suffered a major setback on the West Coast.

In contrast, Continental has rapidly expanded its Kayo brand, nearly tripling the number of stations over a ten year period. If Continental

TABLE 8-2

MAJOR OIL COMPANIES' SECONDARY BRANDS

Major Oil Company	Date Acquired	Number Stations Purchased	Brand	Number Stations 1970	Comments
Gulf	1960	600	Wilshire	N/A	971 stations converted to Gulf in 1965
Union			Harbor		Gradually selling off its stations
Continental	1961	210	Douglas	526	Sold its locations in L.A. to Texaco
Continental	1959	170	Kayo	484	
Continental	1959	300+ (est.)	Western Fuel Oil Direct Mileage Western	56	Direct & Mileage now have new Conoco trademark
Sun (Sunray DX)			Premier	700	
Sun (Sunray DX)			Freeway		Conversion of unsuccessful DX stations
Union (Pure)	1959	30 (est.)	Wisco		Sold 21 stations to Martin Oil Service in 1969
Atlantic-Richfield			Rocket		
Phillips			Seaside	336	
Tenneco	1962	132	Direct	178	

continues to expand the Kayo brand and operates as a pure marketer, this particular merger would not seem to have had a detrimental effect on intertype competition. However, it remains to be seen what Continental will do with the Kayo stations over the long run and if Kayo will eventually go the way of Douglas, Direct, and Mileage. The same questions need to be raised about the future of the Premier stations acquired by DX before its merger with Sun. Sun has been outspokenly opposed to the price marketers' method of operating and claims to not generally sell unbranded gasoline. If Premier, as a result of the merger becomes less aggressive or converts to Sun as DX has done, intertype competition will again be weakened.

A question frequently raised about the majors' secondary brands has been, "Are they bogus companies being used for anticompetitive purposes or legitimate business propositions?" One of the characteristics of the old Standard Oil Trust was that some of the independent oil companies they purchased were operated at a loss for a period of time for anticompetitive purposes. There is strong feeling and some evidence to suggest that secondary brands have been used to regulate genuine price marketers; that is, that they have been used as fighting brands and in areas where price marketers are concentrated.

MERGER OF PRICE MARKETERS INTO INTEGRATED INDEPENDENTS

THE ONLY SEGMENT OF THE independent industry that grew during the price-war years was the integrated independents (see Chapter 6, pp. 177–78). A large part of their growth was achieved by mergers with price marketers and by contractual integration with private branders who agreed to sell refiner brands in order to get price protection. Four of the important integrated independents that grew by such mergers were Murphy, Hess, American Petrofina, and Signal Oil and Gas.

Murphy Oil Company tripled its retail outlets through mergers with Spur in 1960 and with Ingram in 1961. Spur was a leading private brander with 348 stations in twenty-one southern and southeastern states at the time of the merger. Much of Spur's appeal was that it generally sold gasoline for two cents less per gallon than the majors and further discounted its price by a strong premium coupon plan.[7] The merger with Ingram, an independent refiner-marketer, brought to Murphy 201 stations that sold gasoline generally for two cents less per gallon than the majors.[8] Under Murphy the pricing policy of Spur and Ingram was changed so that by the middle 1960's many of the stations were pricing gasoline at the majors' level. This was accomplished by an upgrading program in which stations were remodeled or rebuilt and by promotion of the Murphy credit-card program. Thus in the case of the Murphy mergers with independents, price competition was weakened in much the same way as in the buy-outs of price marketers by majors and their conversion to the major brand.

Hess Oil Company also was an active acquirer of private branders and refiner-marketers in the latter 1950's and early 1960's. In 1958 Hess purchased 25 percent of the Meadville Corporation, the largest East Coast private brander with 250 stations selling the Merritt, Save Way, Safe Way, and Giant brands. In 1962 Hess increased its holdings in

Meadville to 49 per cent. Hess also purchased in 1963 Billups Eastern with 170 private-brand stations in the East and Southeast. The Meadville Corporation continues as an aggressive East Coast private brander. The Billups Eastern operation was upgraded and converted to the Hess Brand which is very price aggressive, selling generally from two to five cents below the majors' prevailing price. As a result of these mergers, and through internal expansion, Hess brand stations have increased from 28 in 1961 to more than 500 in 1969. Its stations average sales of about 100,000 gallons a month—the highest station average for a chain of its size.[9] Such gallonage is achieved by a combination of factors including aggressive pricing, good locations, outstanding appearing stations, and excellent service. So far, the mergers culminated by Hess appear to have thus far strengthened price competition on the East Coast.

However, what remains to be seen is whether Hess will continue over the long run to be a strong price marketer. Some of those interviewed for this study have indicated that they felt eventually the major oil companies would compel Hess to market on a nonprice competitive basis. Supposedly, Hess has concentrated volume in three geographic areas where other majors could lower prices and exert pressure on it to cease its aggressive pricing practices. Furthermore, in appraising Hess's mergers with independents, it should be recognized that they have contributed to the increasing vertical integration taking place throughout the industry. The Hess brand stations and others owned partially by Hess are controlled outlets of a vertically integrated system.

American Petrofina, starting from scratch in the middle 1950's, grew rapidly by a series of mergers of smaller independents. By the early 1960's the amalgamated operation had over 3,000 stations selling the Fina brand throughout seventeen mid-continental states—the largest branded station representation of the integrated independents selling price-discounted gasoline. Fina's pricing policy seems generally to be two cents under the prevailing major-brand price. While Fina's strategy is still based upon discounted price, it has gone in for some of the marketing frills—brand advertising, credit cards, and fancy stations—which make it somewhat less price aggressive. Frequently, other private branders will price a penny under the Fina brand or give premiums or stamps worth another penny per gallon. There are some indications that American Petrofina may be more inclined to go along with established price differentials and be more interested in the status quo relationships than other integrated independents.

Another large integrated independent that grew by successive mergers of independents in the latter 1950's and early 1960's was Signal Oil and Gas. When Socal elected not to renew its contract to buy Signal's crude-oil production in 1957, Signal had two choices—either to sell out or to integrate forward. Signal elected to integrate forward in order to sell its crude-oil production of approximately 55,000-barrels-a-day. Signal Oil and Gas merged with independent refiners—Hancock, the largest West Coast independent refiner in 1958, and Bankline and Eastern States in 1959. Signal then completed a number of mergers with private branders that gave it 1,000 stations in five western states under five brands—Hancock, Bankline, Century, Regal, and Western Hyway. In addition, Signal acquired a controlling interest in Billups Western and Supertest, two large private branders with extensive operations in the Southeast.

The Signal mergers of several significant independents apparently have not worked and Signal is in the process of selling off the acquired assets. In 1970 Signal sold its 70,000-barrel-a-day Houston refinery and 914 stations (Hancock, Billups Western, and Supertest) in the Southeast to a Florida holding company. In another transaction Signal sold its 22,000-barrel-a-day Bakersville, California, refinery in 1970.[10] The remaining 1,061 stations in the five far western states are being sold to a number of buyers. Presently it appears that several hundred Signal stations in the western states will be converted to major brands. However, the final disposition of the approximately 2,000 stations is unclear at this time. Should they ultimately be sold to majors, or converted to the majors' competitive style, intertype competition will suffer still another major setback.

Summary and Conclusion

The late 1950's and early 1960's witnessed buy-outs and sell-outs of most of the significant price marketing chains (also of those private branders who did not price aggressively). Major integrated oil companies such as Standard Oil of New Jersey, Gulf, and Continental and large integrated independents such as Hess, Signal Oil and Gas, American Petrofina, and Murphy purchased chains of aggressive price marketers with thousands of stations throughout the country. The push for controlled gallonage literally wiped out a large proportion of the independent marketing chains and carried the oil industry a big step closer to complete vertical integration.

There seemed to be two primary reasons for the buy-outs of the price-marketing chains. First mergers beget more mergers. Out of self-defense a supplier who has lost customers as a result of mergers by others is likely to look for merger partners. Since some of the first significant mergers, like those made by Standard Oil of New Jersey in the latter 1950's, were not contested, they tended to act as catalysts for others which quickly followed.

A second reason for the rash of mergers involving the independents was the turbulent price conditions in certain parts of the country that were discussed in Chapters 5 and 6. The depressed profits and operating losses of many companies and the uncertainty about whether conditions would ever improve encouraged or forced many mergers. Most of the independents were taken over by bigger and more diversified operators who were in better positions to withstand price wars and margin squeezes affecting refiners and marketers of gasoline. For example, those companies with sizable crude-oil production had sheltered and protected profits and those companies with widespread operations had the advantages of regional diversity.

Many of the mergers involving price marketers have clearly reduced intertype competition, while others seem to have had either a neutral effect or a mildly positive effect. Those that clearly reduced intertype competition were the purchases of price marketers by major oil companies (and integrated independents) who converted them to major brands and nonprice marketing methods. Examples of such mergers involved Standard Oil of New Jersey, Gulf, and Murphy. In Chicago and much of the West Coast such mergers have definitely weakened intertype competition. On the other hand, the purchases of price marketers by some of the major oil companies such as Continental and Tenneco, and their continued operation as aggressive secondary price brands, does not yet seem to have weakened intertype competition. However, there are some difficult and long-run questions associated with this practice that still remain to be answered. For example:

1. Will the integrated independents ultimately be converted to the major brand?
2. Are they used primarily as fighting brands for anticompetitive purposes?
3. Do they result in discrimination against the branded dealer?

In addition, these mergers have contributed to increased vertical integration with its attendant problems.

The growth of the integrated independents through mergers of smaller independents does not in many instances seem to have reduced intertype competition. The integrated independent, because of its vertical integration and regional diversity, seems to have the scale of operation necessary to compete effectively with the major oil companies. This would include operations such as Hess Oil and Chemical in the East and American Petrofina in the Mid-continent. However, other integrated independents have not been successful. Signal Oil and Gas is in the process of selling its refineries and stations and the effect this will have on intertype competition is still to be determined. While the results are mixed, when the balance is cast, the buy-outs over the past fifteen years have harmed intertype competition in the main.

9 / *Vertical Integration and Monopoly Power*

THE MATERIALS PRESENTED TO THIS POINT have principally described and analyzed the nature of competition in gasoline marketing without considering gasoline marketing in the context of the broader petroleum industry. Now, however, it is necessary to broaden the analysis since the way in which gasoline is marketed is greatly influenced by the structure of the petroleum industry. Furthermore, should the performance of the gasoline industry be unsatisfactory in certain respects, public-policy reforms are more likely to be effective over the long run if they are directed at structural reform in the basic industry rather than at strategic aberrations.

In this chapter the important structural features of the petroleum industry which seem to govern the strategies used in marketing gasoline will be explored. In addition, technological, political, economic, and legal factors will be considered which seem to be shaping the structure and the strategy of the industry.

Basic Structural Characteristics

IN THE GASOLINE BUSINESS two important structural features stand out. First, each geographic gasoline market is dominated by a relatively small number of financially powerful sellers, who also operate throughout different parts of the country. This structural property is called oligopoly. Second, frequently the successive stages of activities in the petroleum industry are owned and coordinated by management of a single company. This structural characteristic is called vertical integration. Both oligopoly and vertical integration have been studied extensively in a variety of different industries and certain strategic patterns repeatedly emerge when these structural properties are present. Thus, being able to classify the petroleum industry as a vertically integrated oligopolistic industry is of considerable assistance in explaining much of the strategic behavior one encounters in the gasoline business. It also means that decisions at one level are influenced by strategic factors present at other levels.

The petroleum industry is dominated by 15 to 20 huge, vertically integrated firms. For example, five of the top ten U.S. firms in total assets are oil companies and the 20 largest oil companies rank among the nation's largest 200 industrial corporations. These oil companies are not only engaged in the marketing of gasoline but also the exploration for crude oil, the production and transportation of crude oil, and the refining of crude oil into gasoline and many other petroleum by-products and derivatives as well. Some information relating to characteristics of the 20 largest integrated oil companies is presented in Table 9-1.

Besides the twenty major integrated oil companies, there are approximately another twenty smaller integrated oil companies classified as semimajors. Because of their smaller-scale operations and the limited impact they have on markets, the semimajors' market behavior frequently parallels that of the independents rather than that of the major oligopolists.

The balance of the activity in the industry is conducted by thousands of smaller nonintegrated or partially integrated companies. Precise figures are hard to come by, but it has been estimated that there are more than 7,000 companies which are engaged in crude-oil production, 150 different refining companies, and more than 10,000 wholesale distributors. These figures of course include the majors and semimajors as well as the independents.

Prevailing Competitive Strategy

THE COMPETITIVE strategy appropriate to such a market structure is relatively easy to predict. The major oil companies will generally refrain from price competition and exhibit leadership in establishing relatively high and stable prices. The economic theory of markets in which there are oligopolistic competitors accurately predicts that the prevailing gasoline price will be set at or near the price posted by one of the dominant sellers. Experience also shows that in any market consisting of

TABLE 9-1

Scale of Operation of the Twenty Largest Integrated Oil Companies

Company	Crude Oil Production (barrels per day)		Refining Capacity (barrels per day)		Branded Stations in United States		Total Assets (millions)	Fortune 500 Asset Rank 1969
	Domestic	Foreign	Domestic	Foreign	Number	States		
Standard Oil of New Jersey	866,000	3,452,000	992,000	3,900,000	29,427	47	$17,538	1
Texaco	866,000	no data	929,000	1,631,000	40,230	51	9,282	3
Gulf	602,000	2,202,000	688,000	765,000	31,271	49	8,105	5
Mobil	350,000	1,090,000	829,000	1,078,000	25,513	45	7,163	7
Standard Oil of California	474,214	1,679,428	688,213	872,242	20,589	34	6,147	10
Standard Oil of Indiana	452,000	193,000	895,000	76,000	29,702	49	5,151	—
Shell	548,000	—	870,000	—	22,000	43	4,356	12
Atlantic-Richfield	454,248	222,729	658,455	41,197	22,778	46	4,235	16
Phillips	267,900	119,200	359,000	75,000	21,296	51	3,102	22
Continental	166,976	294,297	261,337	27,009	6,900	30	2,897	24
Sun	295,000	113,000	434,000	36,000	16,900	38	2,528	28
Union	288,600	51,400	392,900	3,900	16,751	41	2,476	36
Cities Service	192,000	11,700	228,000	26,000	9,459	30	2,066	58
Getty	299,000	89,000	195,000	56,000	2,513	11	1,859	42
Standard Oil of Ohio	28,000	23,000	351,000	—	12,800	18	1,554	58
Marathon	156,520	270,233	165,334	34,961	3,615	9	1,300	75
Amerada Hess	91,566	142,162	291,000	—	no data	14	982	103
Ashland	13,811	14,510	255,138	—	2,813	10	846	117
Kerr McGee	30,702	2,576	47,330	—	1,835	18	668	139
Skelly	84,681	4,093	64,660	—	4,350	17	648	—

Source: National Petroleum News Factbook Issue (*Mid-May 1969*), *pp. 30–31.*

a small number of dominant sellers, the implicit understanding that a price cut can and will be met by all other large sellers, effectively serves to deter frequent or aggressive price competition. When price competition is unlikely to offer any major seller additional profit for long, competition for a share of the market is diverted to the nonprice forms which have come to characterize the major marketing style described in Chapter 2. Price competition, when it occurs, will produce the destructive price warfare described in Chapters 5 and 6.

Effect of Vertical Integration

While the gasoline market does tend to behave like a classic oligopoly, the petroleum industry is not nearly as concentrated as many other industries which seem to be workably competitive. There are probably just too many gasoline sellers, and the geographic distribution of their shares of the market is too uneven for price competition to be completely suppressed as it is in those oligopolistic industries dominated by fewer than one-half dozen sellers. Without the presence of vertical integration, there is reason to believe that at least the retailing of gasoline could be workably competitive from a practical standpoint. Competitive self-restraint among twenty to thirty firms is likely to breakdown as each jockeys for competitive advantage. Even sporadic price competition among this many sellers would serve to keep prices and margins in line, rewarding the innovators and the efficient and disciplining the laggards and the inefficient. However, vertical integration is very much present in the industry, and its presence effectively frustrates this potentially workable competition at retail for a number of reasons. Because it frustrates retail competition, vertical integration strengthens the tendency of a marginally oligopolistic market to behave as a classic textbook example of oligopoly at its worst.

In many ways, then, it is vertical integration rather than the scarcity of sellers which colors the marketing strategies followed by the major firms in the industry. Without vertical integration, many features of the major-brand style of marketing wouldn't make any sense at all. The top-heavy investment in excessive retail facilities would collapse. Resistance to marketing innovations would relax. Also, without vertical integration variety in the retail offer would flourish, and motorists would at last have the chance to choose the particular combination of product image, ancillary service, and price that they want and for which they are willing to pay.

It is interesting that vertical integration in this industry has this effect, for on the surface at least, vertical integration as such doesn't appear to be especially anticompetitive. In theory, if an industry is competitive at each level, the vertically integrated firm would not be able to gain a strategic advantage by concentrating profits at one level and subsidizing other activities to squeeze its nonintegrated competitors. Excess profits at any one level would invite new investment and the attendant expansion of that activity would quickly drive excessive returns at that level back down to normal. This also means that no firm could long accept subnormal returns at any one level in hopes of recouping at another, since recoupment would be impossible if returns at all levels were competitively determined.

Monopoly Power in Different Industry Activities

If any level of the industry is not competitive, then it is possible for vertical integration to confer a strategic advantage on the vertically integrated firm. In fact, the less competitive a particular level of industry activity, the greater the likelihood that a vertically integrated firm will use its market power at that stage to gain a competitive edge on its nonintegrated competitors at other levels. This is clearly the case in the petroleum industry where the state-abetted quasi-monopolistic position of the major oil companies at the crude-oil level has been extended by vertical integration to the potentially competitive refining and marketing levels.

Early Dominance in Refining

In the past, market power at the refining and pipeline levels was extended to other competitive levels in this way. John D. Rockefeller's original Standard Oil empire was founded on the monopolistic control of intermediate markets. Through a syndicate of some thirty-three companies the Standard Oil Trust achieved a position of dominance at the refining level (controlling 85 percent of refining capacity at one point) which was then effectively protected by integrating backwards into transportation. As a result of its dominance in refining and its position in transportation, the Trust was able to effectively control the entire industry at all levels.[1]

Evidence of the strategic advantage that vertical integration can confer was demonstrated by the fact that most of the new firms created by the dissolution of the Standard Oil Trust in 1911 quickly moved to integrate forward and backward if they were not already so integrated. In fact, they were more or less compelled to do so in order to protect themselves against their vertically integrated competitors.

Dominance in Pipelines

WITH THE break-up of the refinery monopoly, the power base from which the vertically integrated firms operated during the next period shifted to pipeline control. Control of the pipelines by the major firms, coupled with the requirement that independent producers sell their oil at the wellhead, put those companies owning pipelines in an advantageous bargaining position vis à vis crude-oil producers. Producers had to accept the integrated buyers' price. Furthermore, by controlling the pipelines, the major integrated firms were able to put the nonintegrated refiners at a competitive disadvantage. The independent refiners either had to pay the high tariff for using the major integrated company-owned pipeline or else confine their operations to the highly competitive markets surrounding the producing area.[2]

The period from 1920 to 1940 was the golden era of the pipelines. The twenty largest integrated oil companies controlled 57.4 percent of crude oil gathering line mileage; 89.0 percent of crude oil trunk line mileage; and 96.1 percent of gasoline line mileage.[3] During this period huge profits were captured by the pipeline affiliates of the major oil companies. For example, during the 1920's the Humble Oil and Refining Company, a subsidiary of Standard Oil of New Jersey, earned from 20 to 33 percent return per year on its pipeline investment, while crude-oil production and refining frequently lost money or contributed only small returns on their investment. Between 1920 and 1930, pipeline profits contributed 85 percent of the company's net earnings while they represented only one-third of the company's total investment.[4]

In 1938, the major integrated oil companies' return on investment (less depreciation) from crude-oil lines was 26 percent and on gasoline lines was 29.7 percent. By contrast, the return on investment (less depreciation) of the independent pipeline companies was only 9.7 percent.

This evidence of excessive earnings from integrated pipelines and the implicit injury to nonintegrated competitors resulted in a major antitrust suit against the pipeline companies in 1940. The suit was ultimately settled by a consent decree in 1942 which limited earnings to 7 percent of valuation of common carrier pipeline assets.[5] While this decree did curtail the capturing of excessive profits in pipelines, the importance of pipelines for major company control of the industry remains.

Dominance in Crude-oil Production

THE LOSS of the profit haven in pipelines could have presented a severe problem to the vertically integrated oil companies. Without the strategic advantages gained from pipeline control, the vertically integrated companies would have faced increasing competition from nonintegrated companies which specialized in the refining, distribution, and the marketing of gasoline. However, a new profit haven for the vertically integrated system was quickly found. As Eugene Rostow speaks of it, the "nerve center of the industry" shifted backward from pipelines to the production of crude oil. Crude oil proved to be an excellent replacement for pipelines as a profit haven for the following reasons:

1. As both large producers and principal buyers the major integrated oil companies were able to administer effectively the price of crude oil.
2. A combination of federal laws permitted the major oil companies and/or producers to balance production with demand at the administered price level for crude oil. (Demand prorationing and import quotas are discussed on pp. 244–52.)
3. Preferential income tax treatment of crude-oil profits makes a dollar of pretax profits at this level more productive of after-tax profits than at other levels. (The preferential tax treatment of crude-oil earnings is discussed on pp. 253–68.)

SHIFTING INDUSTRY PROFITS BACKWARD INTO CRUDE OIL

THE SHIFTING OF INDUSTRY PROFITS backward into the crude-oil department gave certain of the large vertical integrated oil companies an important competitive advantage over the independents—

refiners, terminal operators, and marketers. In addition, the high administered crude-oil prices also worked to the disadvantage of the weaker integrated companies having to purchase a large portion of their crude oil from others at artificially inflated prices. The gradual shifting of industry profits into crude-oil production manifested itself in a variety of forms: departmental changes in company profits; changes in earnings reports of integrated and nonintegrated companies; major shifts in revenue going to industry specialists; and changing investment policies.

The shifting of petroleum industry profits into crude-oil production following the enactment of the conservation laws passed in the middle 1930's is illustrated by the operating results of the Humble Oil and Refining Company. As noted earlier, between 1919 and 1930, 85 percent of Humble's net earnings were captured by its pipeline division which accounted for only approximately one-third of the company's assets. The high point was reached during the period 1930 to 1933 when Humble's pipeline division contributed more than 100 percent of the company's total profits. After 1933, as the oil conservation laws began

TABLE 9-2

HUMBLE'S PIPELINE PROFIT RELATIVE TO TOTAL PROFITS FROM ALL ACTIVITIES

Year	Humble's Total Earnings (Add 000)	Humble's Pipeline Earnings (Add 000)	Pipeline Earning as % of Total
1931	$ 2,765	$22,150	800
1932	14,897	15,739	105
1933	20,848	10,152	49
1934	21,990	7,676	35
1935	23,966	6,746	28
1936	34,184	7,957	23
1937	46,924	11,713	25
1938	35,800	9,283	25
1939	29,950	7,582	25
1940	28,108	5,892	21
1941	35,357	6,518	18
1942	29,243	1,920	7
1943	45,712	4,131	9
1944	60,562	6,367	11
1945	70,895	3,266	5
1946	71,832	4,167	6
1947	124,107	3,164	3
1948	186,069	8,821	5

Source: Henrietta M. Larson and Kenneth W. Porter, History of Humble Oil and Refining Company *(New York: Harper and Bros., 1959), pp. 692 and 698.*

to take hold, pipeline profits steadily declined in importance (see Table 9-2), and by the war they were completely eclipsed by crude-oil profits which accounted for three-fourths of Humble's pretax profits during the 1942 to 1945 period.[6]

Effect on Crude-oil Production Companies

As a result of the major vertically integrated oil companies' decision to capture large industry profits at the crude-oil production level one might expect them to share the bounty of higher administered crude-oil prices with companies specializing in crude-oil production. If this were the case, one would expect to find crude-oil production companies having a higher return on investments than the vertically integrated oil companies who subsidize downstream refining, distribution, and marketing operations through crude-oil profits. This is in fact what one finds when comparing the return on investments of several specialized crude-oil production companies with vertically integrated oil companies (see Table 9-3). From 1942 on, those companies specializing in crude-oil production have consistently earned a higher return than the vertically integrated companies since they don't have the subsidized downstream operations. The difference in the earnings of the crude-oil specialist and the integrated oil companies became particularly pronounced following World War II when crude-oil prices doubled over a two-year period.

The return on investment data shows another characteristic of the industry—namely the large amount of wealth concentrated in the hands of the vertically integrated companies. Through the thirty-two-year period the integrated oil companies have commanded ten to twenty times the assets of the crude-oil producing companies. Had the producing companies not been relatively small, the integrated oil companies would have been much less willing to share the bounty from the high administered price of crude oil with them.

Effect on Independent Refineries and Other Independents

While the strategy of the vertically integrated oil companies to capture disproportionately high industry profits in crude-oil production inflated the returns of the independent oil producers, the opposite could be expected for the independent refiners having to pay a high

TABLE 9-3

Return on Investment of Production Companies and Vertically Integrated Firms from 1937 to 1968
(thousands of dollars)

	NUMBER OF COMPANIES	NET INCOME AFTER TAXES	BOOK NET ASSETS, JANUARY 1	PERCENT RETURN ON NET ASSETS
1968:				
Oil and gas producing	61	369,923	2,310,281	16.0
Integrated operations	38	5,757,891	45,233,757	12.7
Total	99	6,127,814	47,544,038	12.9
1967:				
Oil and gas producing	65	327,380	2,054,730	15.9
Integrated operations	42	5,368,379	42,156,722	12.7
Total	107	5,695,759	44,211,452	12.9
1966:				
Oil and gas producing	61	291,379	1,914,132	15.2
Integrated operations	45	4,883,249	39,139,696	12.5
Total	106	5,174,628	41,053,828	12.6
1965:				
Oil and gas producing	62	268,350	1,810,700	14.8
Integrated operations	47	4,369,179	37,000,552	11.8
Total	109	4,637,529	38,811,252	11.9
1964:				
Oil and gas producing	73	254,021	1,568,754	16.2
Integrated operations	49	3,985,479	35,015,397	11.4
Total	122	4,239,500	36,584,151	11.6
1963:				
Oil and gas producing	69	177,314	1,345,960	13.2
Integrated operations	46	3,742,867	33,065,503	11.3
Total	115	3,920,181	34,411,463	11.4
1962:				
Oil and gas producing	80	208,655	1,694,774	12.3
Integrated operations	42	3,089,821	29,648,615	10.4
Total	122	3,298,476	31,343,389	10.5
1961:				
Oil and gas producing	80	206,567	1,628,910	12.7
Integrated operations	44	3,018,597	29,386,129	10.3
Total	124	3,225,164	31,015,039	10.4
1960:				
Oil and gas producing	82	189,291	1,616,186	11.7
Integrated operations	43	2,834,818	28,016,654	10.1
Total	125	3,024,109	29,632,840	10.2
1959:				
Oil and gas producing	88	192,377	1,571,018	12.2
Integrated operations	43	2,621,957	26,564,949	9.9
Total	131	2,814,334	28,135,967	10.0
1958:				
Oil and gas producing	76	199,991	1,688,244	11.8
Integrated operations	45	2,397,492	23,711,808	10.1
Total	121	2,597,483	25,400,052	10.2

TABLE 9-3 (continued)

	NUMBER OF COMPANIES	NET INCOME AFTER TAXES	BOOK NET ASSETS, JANUARY 1	PERCENT RETURN ON NET ASSETS
1957:				
Oil and gas producing	71	219,290	1,331,796	16.5
Integrated operations	45	3,020,272	22,480,202	13.4
Total	116	3,239,562	23,811,998	13.6
1956:				
Oil and gas producing	58	182,433	1,138,438	16.0
Integrated operations	43	2,962,389	20,313,065	14.6
Total	101	3,144,822	21,451,503	14.7
1955:				
Oil and gas producing	50	148,611	916,241	16.2
Integrated operations	42	2,621,941	18,572,736	14.1
Total	92	2,770,552	19,488,977	14.2
1954:				
Oil and gas producing	51	190,100	1,243,333	15.3
Integrated operations	43	2,266,133	16,487,795	13.7
Total	94	2,456,233	17,731,128	13.9
1953:				
Oil and gas producing	52	174,262	1,060,744	16.4
Integrated operations	43	2,187,878	15,341,547	14.3
Total	95	2,362,140	16,402,291	14.4
1952:				
Oil and gas producing	48	155,698	955,853	16.3
Integrated operations	43	2,016,707	14,043,139	14.4
Total	91	2,172,405	14,998,992	14.5
1951:				
Oil and gas producing	46	159,157	816,997	19.5
Integrated operations	45	2,102,871	12,715,442	16.5
Total	91	2,262,028	13,532,439	16.7
1950:				
Oil and gas producing	36	110,004[1]	600,527	18.3
Integrated operations	45	1,730,484	11,618,635	14.9
Total	81	1,840,488	12,219,162	15.1
1949:				
Oil and gas producing	40	107,888[1]	496,154	21.7
Integrated operations	44	1,420,689	10,761,367	13.2
Total	84	1,528,577	11,257,521	13.6
1948:				
Oil and gas producing	44	143,751[1]	400,051	35.9
Integrated operations	44	1,954,277	8,844,742	22.1
Total	88	2,098,028	9,244,793	22.7
1947:				
Oil and gas producing	41	69,983[1]	351,980	19.9
Integrated operations	40	1,215,947	7,712,538	15.8
Total	81	1,285,930	8,064,518	15.9
1946:				
Oil and gas producing	44	36,504[1]	289,098	12.6
Integrated operations	40	760,592	7,092,034	10.7
Total	84	797,096	7,381,123	10.8

TABLE 9-3 (continued)

	NUMBER OF COMPANIES	NET INCOME AFTER TAXES	BOOK NET ASSETS, JANUARY 1	PERCENT RETURN ON NET ASSETS
1945:				
Oil and gas producing	40	35,379[1]	272,110	13.2
Integrated operations	37	447,778	5,322,478	8.4
Total	77	483,157	5,594,588	8.6
1944:				
Oil and gas producing	43	30,911[1]	247,882	12.5
Integrated operations	39	624,922	6,443,010	9.7
Total	82	655,833	6,690,892	9.8
1943:				
Oil and gas producing	40	24,381[1]	225,774	10.8
Integrated operations	37	289,623	3,650,662	7.9
Total	77	313,004	3,876,436	8.1
1942:				
Oil and gas producing	46	24,594[1]	272,682	9.0
Integrated operations	35	258,961	3,693,152	7.0
Total	81	283,555	3,965,834	7.1
1941:				
Oil and gas producing	44	22,937[1]	275,537	8.3
Integrated operations	34	462,431	4,655,316	9.9
Total	78	485,368	4,930,853	9.8
1940:				
Oil and gas producing	42	12,740[1]	297,823	4.3
Integrated operations	40	251,955	4,689,854	5.4
Total	82	264,695	4,987,677	5.3
1939:				
Oil and gas producing	41	13,266[1]	275,636	4.8
Integrated operations	39	222,493	4,126,218	5.4
Total	80	235,759	4,401,854	5.4
1938:				
Oil and gas producing	46	15,352[1]	234,745	6.5
Integrated operations	47	250,888	4,972,115	5.0
Total	93	266,240	5,206,860	5.1
1937:				
Oil and gas producing	46	20,522[1]	224,228	9.2
Integrated operations	47	471,299	4,695,445	10.0
Total	93	491,821	4,919,673	10.0

1. Before depletion in some cases.
Source: U.S., Congress, Senate, Committee of the Judiciary, Subcommittee on Antitrust and Monopoly, First National City Bank of New York City, reported in Government Intervention in the Market Mechanism—The Petroleum Industry, *Part 2,* Industry Views, *91st Cong., 1st sess., pp. 1063–65.*

price for their basic raw material. Unfortunately, there are no published accounts of the history of refining companies which are comparable to those for vertically integrated companies and crude-oil companies.

However, there is information on refining spread and costs relative to product prices which is summarized in Tables 9-4 and 9-5. Note that while gasoline prices have risen from $4.56 per barrel in 1950 to $4.99 per barrel in 1970 (Col. b), crude-oil prices have risen from $2.52 per barrel to $3.13 per barrel over the same period (Col. a). However, since refining realizations are a function of the relative yield of different products from a barrel of crude oil and the prices of these different products, comparisons of crude-oil and gasoline prices alone can be misleading. A more accurate picture of profit trends in refining can be gleaned by comparing the trend in crude-oil prices (Col. a) with the average price of the four major refined products weighted by their relative yield (Col. c). The actual procedure is explained in the footnote at the bottom of Table 9-4.

In Table 9-4, average crude-oil prices in eight refining markets (Col. a) are subtracted from the weighted average of four product prices in these same eight markets (Col. c) to get the refining spread (Col. d). Average total operating costs for U.S. refineries (Col. e) are then subtracted from this refining spread to get a rough approximation of the trend in refining profits (Col. f). Note how the squeeze on refining operations has produced successive losses since the mid-1950's.

The shifting of industry profit from refining to the crude-oil department has quite naturally had a devastating impact on independent refining operations. From 1938 to 1970 the share of refining capacity accounted for by the twenty largest refiners had increased from 79.5 percent to 85.7 percent while the share of smaller nonintegrated refiners has declined from 21.5 percent to 14.3 percent.[7] Furthermore, most of the remaining 14.3 percent are now small integrated companies themselves. With the artificial inflation of crude-oil prices it became practically impossible to operate an independent refinery without the refinery producing a significant proportion of its own crude-oil requirements. The principal exceptions are those independent refineries operating in the northern tier of the country or on the West Coast that are able to reduce their crude-oil costs by processing a high proportion of low-price imported crude oil and a few refineries competing in relatively sheltered markets. Dropped from the ranks of the independent refiners during the 1950's and 1960's were the Wood River Refinery (sold to

TABLE 9-4

Changes in Crude-oil Prices, Product Prices, and Refining Costs, 1935–1970

Eight Markets

Year	Crude-oil Prices $/Bbl. (a)	Gasoline Prices $/Bbl. (b)	4 Products Average $/Bbl. (c)	Spread $/Bbl. (d)	Av. Total Ref. Cost $/Bbl. (e)	Av. Profit $/Bbl. (f)
1970	3.132	4.99	4.16	1.028	NA	—
1969	3.077	4.96	3.89	.813	1.370	−.557
1968	2.941	4.85	3.84	.899	1.347	−.448
1967	2.915	4.97	3.92	1.005	1.337	−.332
1966	2.882	4.87	3.84	.958	1.277	−.319
1965	2.864	4.84	3.83	.97	1.225	−.259
1964	2.877	4.73	3.71	.83	1.215	−.382
1963	2.894	4.77	3.79	.90	1.243	−.347
1962	2.905	4.84	3.84	.94	1.247	−.312
1961	2.886	4.88	3.87	.98	1.274	−.290
1960	2.882	4.88	3.84	.96	1.271	−.313
1959	2.903	4.89	3.87	.97	1.259	−.289
1958	3.013	4.93	3.89	.88	1.261	−.381
1957	3.087	5.18	4.24	1.15	1.256	−.106
1956	2.788	4.94	3.96	1.17	1.187	−.017
1955	2.765	4.88	3.81	1.05	1.083	−.038
1954	2.775	4.90	3.73	.96	1.111	−.151
1953	2.684	5.04	3.76	1.08	1.101	−.021
1952	2.526	4.78	3.62	1.09	1.015	.075
1951	2.532	4.79	3.69	1.16	.966	.194
1950	2.515	4.56	3.48	.97	.900	.070
1949	2.55	4.50	3.34	.79	.868	−.078
1948	2.59	4.42	3.70	1.11	.898	.212
1947	1.92	3.60	2.91	.99	.811	.179
1946	1.39	2.77	2.21	.82	.746	.074
1945	1.20	2.72	2.07	.87	.674	.196
1940	1.02	2.21	1.64	.62	.510	.110
1935	.97	2.28	1.65	.68	.416	.264

Source: Data for Columns a, b, c, d from Platt's Oil Price Handbook and Oilmanual, *45th Edition (New York: McGraw-Hill, 1969), p. 194. Data for Column e from* Petroleum Facts and Figures, *1971 Edition (New York: American Petroleum Institute, 1971), p. 117. Data for 1969 and 1970 from* National Petroleum News Factbook, *Mid-May 1971 Edition (New York: McGraw-Hill, 1971), p. 108.*

All prices based on low quotations from *Platt's Oilgram Price Service.*

IPAA's figures on wholesale prices of crude petroleum and principal products show trends in oil prices. They were prepared to provide a "price index" similar to but more comprehensive and informative than data published by U.S. Bureau of Labor Statistics. Average prices should not be interpreted as showing the actual realization for producers or refiners during a particular period. The difference between crude and product prices reflects comparative changes in these prices, but is not a measure of refinery margins which involve transportation and refining costs as well as changes in refinery yields.

WEIGHTS

U.S. Including California: Individual product prices weighted as follows: Okla. (16.8%), Midwestern Group 3 (20.8%), New York Harbor (11.2%), Philadelphia (4.0%), Jacksonville (2.4%), Boston (2.4%), Gulf Coast, (22.4%), and Los Angeles (20%). Four products average weighted as follows: Gasoline 50%, Kerosine 5%, Distillate 15% and Residual fuel 30%.

Sinclair); Frontier Refinery (now Ashland); Panhandle Oil (American Petrofina); American Liberty (American Petrofina); Aurora (now Marathon); Malco (Conoco); Bay Refining (Tenneco); Hancock (Signal Oil and Gas); Elk (Pennoil United); Eldorado Refining (American Petrofina); Cosden (American Petrofina); Delphi Taylor (Hess Oil and Chemical); Frontier (Husky); Premier (Sun Oil); Leonard (Total Petroleum); Northwestern (Ashland); and Sequoia (Conoco).

Backward Integration of Vertically Integrated Oil Companies

ANOTHER significant consequence of this strategy of capturing industry profits at the crude-oil level is the pressure it puts even on the vertically integrated oil companies to increase their position in crude-oil production. Vertically integrated companies that want to prosper have to produce a very high proportion of the crude-oil requirements of their own refineries—that is, they have to have a high crude-oil self-sufficiency ratio. Backward integration can readily be observed by studying the efforts of vertically integrated companies with low crude-oil self-sufficiency ratios that have sought to improve their crude position following the implementation of conservation laws.

One of the most notable cases of backward integration involves the Humble Oil and Refining Company. A clear indicator of Humble's tremendous drive for crude-oil reserves is the shift in investment policy which began in the early 1930's. During the period 1917 to 1930, the Golden Age of the pipelines, Humble invested 52 percent of its funds in production, 25 percent in pipelines, and 22 percent in refining and marketing (see Table 9-6). During the decade of the 1930's, as Humble made its big push to build crude-oil reserves, the relative investment in crude-oil production was increased to 78 percent of the total. For the period from 1941 to 1948, Humble's investment in crude-oil production still amounted to 75 percent of its total investment. In the process of changing its investment policy, Humble's share of U.S. crude-oil reserves increased from an average of 1.1 percent from 1926 to 1930, to 8.2 percent from 1931 to 1935, to finally 12.9 percent from 1936 to 1940 (see Table 9-7). Humble thus emerged in 1940 as the company with the largest U.S. crude-oil reserves—14 percent of the total—and substantially strengthened the domestic crude-oil sufficiency ratio of its parent company—Standard Oil of New Jersey.

TABLE 9-5

AVERAGE OPERATING COSTS OF U.S. REFINERIES, 1935 to 1969

(Cents Per Barrel)

Year	Purchased Fuel	Total Labor	Purchased Power	TEL, Chemicals, and Supplies	Maintenance Materials	Insurance and Taxes	Royalties or Research	Obsolescence and Improvements	Interest on Capitalization	Total Costs
1969[1]	15.8	47.7	3.8	24.6	7.6	5.4	9.2	11.7	11.2	137.0
1968	15.5	46.1	3.9	25.5	7.3	5.5	7.7	11.8	11.4	134.7
1967	16.3	46.3	3.8	26.4	7.1	5.2	6.3	11.4	10.9	133.7
1966	14.5	44.0	3.3	26.8	7.0	5.2	4.8	11.3	10.8	127.7
1965	13.2	44.3	3.5	24.5	6.9	5.1	4.6	11.0	9.4	122.5
1964	12.0	44.7	3.5	23.2	6.9	5.0	4.3	12.1	9.8	121.5
1963	11.1	45.3	3.4	21.8	6.8	5.0	4.1	15.1	11.7	124.3
1962	9.2	46.0	3.4	19.8	6.6	4.6	4.3	17.3	13.5	124.7
1961	8.8	49.0	3.4	21.2	6.4	4.5	3.9	17.0	13.2	127.4
1960	6.7	50.3	3.1	22.9	6.0	4.5	3.8	17.8	12.0	127.1
1959	6.0	51.2	2.8	22.0	6.7	4.4	3.8	17.3	11.7	125.9
1958	5.7	52.7	2.7	21.4	6.7	4.2	3.5	17.9	11.3	126.1
1957	7.3	53.2	2.6	20.9	6.5	4.0	3.2	17.1	10.8	125.6
1956	6.0	50.2	2.1	20.5	6.6	3.8	3.0	16.2	10.3	118.7
1955	5.3	45.5	1.9	17.9	6.7	3.6	2.6	15.3	9.5	108.3
1954	5.3	45.7	1.7	16.8	6.8	3.3	2.5	20.2	8.8	111.1
1953	6.3	47.0	1.6	16.1	6.7	3.1	2.4	18.7	8.2	110.1
1952	6.1	44.6	1.5	14.3	6.3	2.8	2.3	16.0	7.6	101.5
1951	6.4	44.2	1.4	13.1	5.9	2.6	2.1	13.9	7.0	96.6
1950	5.7	41.0	1.4	12.1	5.7	2.5	2.0	12.9	6.7	90.0
1949	5.5	38.6	1.3	10.4	5.6	2.4	1.9[2]	14.7	6.4	86.8
1948	8.4	42.6	1.2	10.4	5.6	2.2	1.7	11.7	6.0	89.8
1947	6.7	38.7	1.3	8.6	5.7	2.1	1.5[2]	10.9	5.6	81.1
1946	5.0	36.7	1.5	7.2	5.5	1.9	1.3	10.2	5.3	74.6

TABLE 9-5 (continued)

AVERAGE OPERATING COSTS OF U.S. REFINERIES, 1935 to 1969

(Cents Per Barrel)

Year	Purchased Fuel	Total Labor	Purchased Power	TEL, Chemicals, and Supplies	Maintenance Materials	Insurance and Taxes	Royalties or Research	Obsolescence and Improvements	Interest on Capitalization	Total Costs
1945	4.4	31.2	1.6	7.4	5.2[2]	1.9	1.1[2]	9.6	5.0	67.4
1944	4.5	28.9	1.5	7.3	4.7[2]	1.8	0.9	13.0	4.8	67.4
1943	4.8	26.1	1.4	6.6	4.1[2]	1.7	1.0[2]	12.4	4.6	62.7
1942	5.0	21.6	1.1	5.4	3.9	1.7	1.0	11.6	4.4	55.7
1941	4.7	18.6	0.9	7.6	3.5	1.6	1.0[2]	11.7	4.2	53.8
1940	4.7	17.6	0.9	6.8	3.4	1.6	1.1	10.8	4.1	51.0
1939	4.8	17.1	1.0	6.0	3.4	1.5	1.0[2]	10.5	3.0	48.3
1938	5.1	17.1	0.9	4.8	3.1	1.5	0.9	10.3	3.0	46.7
1937	5.8	16.1	1.0	4.5	3.0	1.5	0.8[2]	9.6	2.9	45.2
1936	5.7	14.1	0.9	4.1	3.0	1.4	0.7	9.4	2.9	42.2
1935	5.6	14.1	1.0	4.0	2.9	1.4	0.6[2]	9.2	2.8	41.6

1. Preliminary.
2. Estimated.

Source: Petroleum Facts and Figures, *1971 Edition* (*New York: American Petroleum Institute*).

TABLE 9-6

CHANGES IN HUMBLE OIL AND REFINING COMPANY'S CAPITAL EXPENDITURES
CAPITAL EXPENDITURE
(IN MILLION DOLLARS)

	Production		Pipelines		Refining and Manu-facturing		Other		Total	
1917–1930	$205.1	52%	$96.7	25%	$86.5	22%	$5.1	1%	$393.4	100%
1931–1940	296.3	78	36.2	9	44.9	12	4.0	1	381.4	100
1941–1948	421.5	75	54.0	10	83.1	15	2.2	—	360.8	100

Source: Compiled from Henrietta M. Larson and Kenneth W. Porter, History of Humble Oil and Refining Company *(New York: Harper and Brothers, 1959), p. 691.*

TABLE 9-7

HUMBLE'S SHARE OF U.S. CRUDE-OIL RESERVES

1926	0.6%	1931	4.5%	1936	12.5%
1927	0.8	1932	7.2	1937	12.3
1928	1.6	1933	7.9	1938	12.4
1929	1.7	1934	9.5	1939	13.2
1930	1.6	1935	11.6	1940	14.0
Average	1.1%		8.2%		12.9%

Source: Henrietta M. Larson and Kenneth W. Porter, History of Humble Oil and Refining *(New York: Harper and Brothers, 1959), p. 394.*

The general move to bolster low crude-oil self-sufficiency ratios following the passage of the conservation laws in the first half of the 1930's is also readily apparent from an analysis of the operating results of Texaco, American, Atlantic, and Sohio (see Table 9-8). Texaco's crude-oil self-sufficiency position increased from 47 percent during the period

TABLE 9-8

CHANGES IN RATIO OF CRUDE-OIL PRODUCTION TO REFINERY RUNS OF SELECT VERTICALLY INTEGRATED OIL COMPANIES

	Texaco	Standard Oil of Indiana	Atlantic	Sohio
1930–1935	47%	26%	23%	0%
1936–1940	60	35	34	1
1941–1945	64	48	55	11
1945–1950	68	50	57	32

Source: John G. McLean and Robert Wm. Haigh, The Growth of the Integrated Oil Companies *(Boston: Graduate Business School, Harvard University), pp. 683–737.*

1930 to 1935 prior to the passage of the conservation laws to 68 percent in the period 1945 to 1950. Since then Texaco's domestic crude-oil position has continued to improve to the point where the company is approximately self-sufficient in crude oil. Texaco's strong crude-oil position has been an extremely important factor in its becoming the largest domestic producer of gasoline in the latter part of the 1960's.

Standard Oil of Indiana approximately doubled its crude-oil self-sufficiency ratio from a low average of 26 percent for the period 1930 to 1935 to an average of 50 percent during the period 1945 to 1950. However, since then Standard of Indiana has not been successful in significantly improving its domestic crude-oil self-sufficiency position and this problem remains a major handicap to the company.

The Atlantic Refining Company's crude-oil self-sufficiency position increased by almost 150 percent from a low average of 23 percent for the period 1930 to 1935 to an average of 57 percent for the period 1945 to 1950. Atlantic continued to build its position by expanding its exploration and production activities and by purchasing several companies with proven reserves. By 1964 Atlantic's combined domestic and foreign production had climbed to 95.5 percent of its refinery runs.[8]

Even Sohio, which produced no crude oil at all during the first half of the 1930's, was producing one-third of its refinery requirements by 1950. Like American, Sohio has been unsuccessful in further improving its crude-oil self-sufficiency position. However, Sohio's recent merger with BP is expected to improve this deficiency and greatly strengthen the company.

The Office of Emergency Preparedness in a 1971 report on the oil industry made the following comment about the pressures resulting in further vertical integration in the oil industry.

> Integrated companies have become *progressively more self-sufficient* in crude production. . . . An increase in crude-oil prices generally will tend to accentuate that trend. It is apparent that the more costly purchased crude becomes, the greater the incentive for integrated companies to expand their own exploration effort (in competition with the independent) and thus reduce their purchases of crude oil in the open market.[9]

Several of the vertically integrated companies with low crude-oil self-sufficiency positions found themselves squeezed between high crude-oil prices and falling market prices during the latter part of the

TABLE 9-9

Changes in Crude-oil Prices, Refinery Margins, and Wholesale Margins

	Price Per Barrel Crude Oil		Refinery Margins				Wholesale Gasoline Margin[1]		
	Amount	Increase from 1951 Price	Per Barrel	Relative to Base Years 51–57			Per Barrel	Relative to Base Years 51–57	
1951	$2.58		$1.07	108%			$1.71	92%	
1952	2.58		.98	99			1.68	89	
1953	2.72	14¢	.94	95			1.75	94	
1954	2.82	24	.82	83	average 100% base period (1951–1957)		2.00	107	average 100% base period (1951–1957)
1955	2.81	23	.95	96			1.99	106	
1956	2.81	23	1.10	111			2.00	107	
1957	3.09	51	1.05	106			1.94	104	
1958	3.03	45	.76	77	83% (1958–1959)		1.93	104	
1959	2.95	38	.88	89			1.90	102	
1960	2.94	37	.90	91			1.84	99	
1961	2.94	37	.93	94		aver-age 86% (1958–1964)	1.72	92	93% (1958–1964)
1962	2.95	38	.87	88			1.64	88	87% (1961–1964)
1963	2.94	37	.82	83	83% (1962–1964)		1.63	88	
1964	2.93	35	.76	77			1.48	80	
1965	2.92	34	.89	90			1.60	91	
1966	2.95	35	.87	88			1.80	97	
1967	2.99	41	.90	91			1.91	102	
1968	3.01	43	.80	81			2.10	112	

Source: Platt's Oil Price Handbook and Oilmanual-*45th Edition* (*New York: McGraw-Hill, 1969*).

1. Difference between the wholesale price of gasoline and dealer tank-wagon price.

1950's and the first half of the 1960's (see Table 9-9). In two crude-oil price advances—one in 1953 and the other in 1957—the price of crude oil increased by approximately fifty cents per barrel. Coupled with the crude-oil price increase of 1957, refinery margins fell in 1958 and 1959 to 83 percent of the average for the previous seven years. After a brief improvement in 1960 and 1961, refinery margins fell once again to 83 percent of the base period 1951 to 1957.

Another indicator of the depressed state of forward markets is the wholesale margin (the difference between refinery price and dealer cost) for gasoline. For the period 1958 to 1964 the wholesale margin for gasoline averaged only 93 percent of the average for the previous seven years. During the worst gasoline price-war years, 1961 to 1964, the wholesale gasoline margin fell to 87 percent of the 1951 to 1957 average.

Mergers of Major Oil Companies Caught in the Price Squeeze

The decline of the forward markets—refining and marketing margins—following the substantial crude-oil price increases of 1957, resulted in a severe and prolonged financial squeeze on all firms with low crude-oil positions. The consequence was inevitable—mergers of companies in an effort to overcome crude-oil deficiency problems. Practically all of the mergers that were completed during the 1960's involved crude-oil poor companies combining with companies having stronger crude-oil positions. A list of seven of the eight principal mergers of the 1960's is presented in Table 9-10.

One might legitimately ask why the crude-rich firms were willing to merge with the crude-poor firms. The motivation of the crude-poor firm is obvious, but the motivation of the crude-rich partner is more complex. In part, the crude-rich firm was motivated by the sheer short-term attractiveness of the financial deal. But in part, it must also have been due to the long-term desire to gain an increasing measure of control over the downstream markets for their crude-oil output.

Failure of some companies to integrate adequately backward. Clearly, one of the consequences of the major oil companies' strategy of capturing industry profits in crude-oil production has been further backward integration. Not all companies have been as successful as Texaco, Standard Oil of New Jersey, and Atlantic in building their crude-oil self-sufficiency position to reduce their vulnerability to increasing crude-oil prices. Those integrated oil companies with relatively low crude-oil self-sufficiency positions could still earn reasonably satisfactory returns so long as the final markets for refined products remained strong. When these markets softened and their prices fell, while crude-oil prices remained high or increased, those firms with low crude-oil self-sufficiency positions experienced a severe financial squeeze. Fred Hartley, president of Union Oil Company of California, explained what happens under those circumstances. "The question is not just one of making more money by pumping rather than buying crude; it is the

TABLE 9-10

CRUDE-OIL SELF-SUFFICIENCY POSITION AND OTHER FACTORS RELATING TO MERGERS OF MAJOR OIL COMPANIES

	Dominant Company in Merger				Secondary Company in Merger				
Year of Merger	Company	Assets (in million dollars)	U.S. Rank	% U.S. Crude-oil Self-sufficiency	Company	Assets (in million dollars)	U.S. Rank	% U.S. Crude-oil Self-sufficiency	U.S. Rank After Merger
1. 1961	Standard Oil of California	2,782	6	72	Standard Oil of Kentucky	144	25	none	5
2. 1965	Union	835	15	74	Pure	679	16	46	10
3. 1965	Atlantic	960	14	60	Richfield	473	20	37	13
4. 1968	Sun	1,412	14	57	Sunray DX	686	17	45	9
5. 1969	Atlantic-Richfield	1,414	11	63	Sinclair	930	16	41	7
6. 1969	Amerada	350	—	—	Hess	486	22	negligible	18
7. 1969	Sohio	819	17	16	British Petroleum				15

crude buyer's vulnerability to getting squeezed between rising crude quotations and falling prices at the gas pump." [10]

Standard Oil of Kentucky was the leading gasoline marketer in the five state area of Kentucky, Georgia, Florida, Alabama, and Mississippi. However, it owned no refineries and had no crude-oil production. It purchased most of its petroleum products from Standard Oil of New Jersey. Basically, a giant gasoline jobber, Standard Oil of Kentucky was obviously quite vulnerable to price changes in gasoline markets. The unsettled market conditions following the crude-oil price increase of 1957 made the prognosis for its future rather bleak. As a result, the 1961 merger with Standard Oil of California made great sense. Socal had strong domestic and foreign production and needed additional marketing facilities to broaden its geographic coverage. Standard Oil of Kentucky had the marketing facilities, but was destined to go nowhere without crude oil.

The effect of this "rising crude price, falling refined product price" squeeze can be observed by studying the profit history of Pure, Richfield, Sunray DX, and Sinclair. As long as market prices remained strong, these crude-oil-deficient companies could pay high prices for crude. This was basically the situation until 1957. However, the subsequent fall in the refining and marketing margins cut deeply into the profits of these companies. From the 1957 level, the average rate of profit the next seven years decreased 13 percent for Pure, 8 percent for Richfield, 31 percent for Sunray, and 37 percent for Sinclair (see Table 9-11). Saddled with high crude-oil prices and low returns for their refining and marketing activities the future for these companies was bleak.

One account of Pure Oil Company's problem explained the company's situation as follows:

> Pure actually loses money on much of the gasoline it sells. "When we charge our refineries with the $3-or-so-a-barrel crude we have to buy on the outside, we lose money," explains Harry Moir, Pure's vice president for marketing.
>
> *Saving Grace.* What saves Pure, of course, is the fact that it does not have to buy all of its crude from the outside. Of the 65 million barrels of crude it refined and sold last year, Pure produced about 27 million from its own wells. On these latter it made good money. "Using our own crude," says Moir, "we make $1 or more a barrel." It was the profits from this crude which accounted for virtually all

TABLE 9-11

PROFIT PERFORMANCE OF VERTICALLY INTEGRATED OIL COMPANIES WITH LOW CRUDE-OIL SELF-SUFFICIENCY POSITIONS
(in thousand dollars)

	Pure	Richfield	Sunray DX	Sinclair
1947	21,197	18,852	—	48,776
1952	27,304	27,225	—	76,825
1956	36,560	—	43,737	91,071
1957	35,524	28,176	57,155	79,308
1958	28,822	20,094	40,664	49,473
1959	28,905	28,058	43,815	45,526
1960	32,555	28,720	41,121	52,547
1961	30,102	25,201	39,932	35,887
1962	28,950	30,615	36,201	47,350
1963	29,767	27,895	41,866	62,704
1964	31,518	21,455	35,182	58,736
1965	merged with Union Oil	merged with Atlantic	38,592	76,673
1966			44,810	94,344
1967			49,172	95,372
1968			merged with Sun Oil	76,583
				merged with Atlantic-Richfield
1957 Profits	35,524	28,176	57,155	79,308
Average Annual Earning 1958–1964	30,088	26,005	39,826	50,318
Percentage of Decline	13	8	31	37

Source: National Petroleum News Factbook *for the years 1948, 1953, 1957–1969.*

of Pure's net income of $23 million last year. In other words, only about $350 million of Pure Oil's nearly $700 million in operating revenues returned any kind of a profit. The rest, in effect, was sheer volume for volume's sake.

"There is no question about it," admits President Milligan. "The growth of our profits is riding on the amount of crude oil we can find." [11]

Pure's poor profit performance after 1957 drew the attention of several groups that maneuvered to take over the company's assets. Early in 1965 Pure's board voted to accept the offer to merge with Union Oil

Company, a strong domestic oil company producing three-fourths of its own crude-oil requirements.

The Richfield merger with Atlantic had much in common with the merger of Pure with Union. In fact, Atlantic had been one of the unsuccessful bidders for Pure's assets. Richfield's profits had stagnated and the company suffered from a low crude-oil self-sufficiency position—it produced only 37 percent of the crude-oil used by its refineries. Atlantic's much stronger crude-oil position, as previously discussed, made the merger a reasonably sound proposition.

Sunray DX was another lackluster profit performer which suffered along with Pure and Richfield as a result of its great dependence upon others for its crude-oil supplies. (It had only 45 percent crude-oil self-sufficiency.) The long-discussed merger with Sun Oil Company was finally completed in 1968. Sunray DX's crude-oil deficiency was strengthened by Sun Oil Company's 57 percent domestic self-sufficiency ratio and its substantial foreign oil production.

The biggest of the mergers of crude-oil-poor integrated oil companies involved Sinclair Oil Company with assets of more than $1.5 billion. Its decade-long poor-profit performance attracted the attention of Gulf and Western, which made a takeover bid for Sinclair's assets late in 1968. However, Sinclair's management decided in favor of a counterproposal by Atlantic-Richfield, which had a domestic crude-oil self-sufficiency ratio of 63 percent, new discoveries in Alaska, and foreign crude-oil production.

In 1969 the Amerada-Hess merger brought together Amerada, a crude-oil production company, and Hess, essentially an East Coast refiner-marketer. Hess had only modest crude-oil production of its own and was forced to pay top dollar for crude oil that it didn't receive under the oil import program. The tie with Amerada gave Hess the crude oil that it very much needed. The following comments give some insight into the logic of the merger:

> The value of having an ample crude-oil supply is demonstrated by a look at Amerada Petroleum and Hess Oil & Chemical. Amerada is a company that produced and sells crude oil. It does not refine it, it does not pump a gallon of gasoline into a car's tank. It has understated assets of $421 million and reserves of crude oil totaling over 600 million barrels in the U.S. and Canada, plus more than this amount in Libya. Hess Oil & Chemical is an aggressive company that has grown from a small, fuel-oil marketer into a $426-million

(assets) company. It has almost no crude oil. Though the book values of their assets are about exactly equal, the market value of Amerada's 12.8 million shares is $1.2 billion while Hess' 10.4 million shares are valued at just $650 million.

When it comes to return on invested capital, including debt, Amerada has the highest of any energy company, while Hess has one of the lowest. Both are special cases, but the comparison points up that the money in oil is in crude.[12]

The last of the big mergers in which crude oil was a major consideration was the joining of Standard Oil of Ohio with British Petroleum-U.S. in 1969. One of the principal reasons why Sohio entered into the merger was its very low domestic crude-oil self-sufficiency position of 16 percent. If British Petroleum's north slope Alaskan oil properties live up to expectation, Sohio's crude-oil self-sufficiency position should be considerably strengthened. In fact, the extent of BP's ownership of Sohio ultimately depends upon the extent of its Alaskan crude-oil production. If production reaches 450,000 barrels a day by 1975, majority control of Sohio will pass to British Petroleum.

Precarious Outlook for a Competitive Supply of Unbranded Gasoline

The strength and even survival of intertype competition is in part dependent upon the availability of an adequate supply of unbranded gasoline at reasonable prices. However, as previously discussed, one of the consequences of administering high crude-oil prices has been the increasing control of the major integrated oil companies over the supply of gasoline. Thus, availability of an economical supply of unbranded gasoline looms as a problem of significant proportion for the price marketers. This problem is not only recognized by the price marketers, but is also a condition apparent to the major oil companies. For example, when discussing the future of the price marketers, a chief executive officer of a major oil company raised the question, "Where are the price marketers going to get their supply over the long-run?"

One of the principal reasons for concern about the continued supply of unbranded gasoline is the forward integration of oil companies into marketing. Through internal growth and mergers the major oil companies have expanded their own marketing operation with the result that a larger portion of the major oil company's gasoline is being sold through their own brand-controlled outlets. The goal of most of the major oil companies, given the industry practice of capturing big profits in crude-oil production with marginal returns from refining and marketing, is perfect balance—produce your own crude oil, process the oil in your own refineries, and sell the gasoline for public consumption through controlled branded outlets. As the goal of balanced integration is approached, the amount of product the major oil companies find necessary and are willing to make available to the unbranded market diminishes.

Another factor reducing the availability of economical sources of unbranded gasoline has been the large number of market-expansion and vertical mergers in the past ten to fifteen years by the major integrated oil companies. Many of the companies taken over by mergers were major sources of supply of unbranded gasoline or outlets for unbranded gasoline. For example, DX which was taken over by Sun, and Sinclair which was principally divided between Atlantic-Richfield and Sohio-BP. were both major suppliers of unbranded gasoline. Concurrent with and since these mergers, DX and Sinclair have been gradually withdrawing from supplying unbranded accounts. In general, such withdrawals from the unbranded markets have taken the form of non-renewals of unbranded supply contracts, increases in prices, and territorial restrictions on supply availability. Clearly companies like Sun Oil that have been outspokenly critical of the price marketing approach are not going to take steps that will foster the growth of the major source of rivalry to the branded nonprice method of marketing gasoline.

Refusal to Deal with Price Marketers

THE REDUCTION in availability of unbranded gasoline from the major oil companies is not only a matter of their forward integration and broader market coverage, but also a basic policy of several of the major integrated oil companies not to sell to price marketers. It is the general practice for the giants of the industry such as Standard Oil of New Jersey, Texaco, Gulf, Mobil, Shell, American, and Union to refuse to supply gasoline to the unbranded market. Instead of letting their gasoline flow into hands which will sell it on a reduced price basis, they reduce refinery runs. While the major oil companies frequently refuse to sell to price marketers, most of them freely exchange and sell gasoline to one another for they are generally insured that their gasoline will not be sold at reduced prices. Furthermore, by exchanging gasoline rather than selling it, open market transactions are few, which decreases the likelihood of excess major-brand gasoline falling into the hands of price marketers.

Role of major oil company controlled finished product pipelines. Refusing to deal with price marketers becomes a very effective device for constraining the nature of competition when it is tied to the ownership of pipelines. For many inland markets the practical and economical way to obtain gasoline is through pipelines. When the pipelines are jointly developed and owned by those majors that have in effect a policy of not selling unbranded gasoline the consequence is obvious—price competition is going to be limited and constrained. For example, the Colonial Pipeline is the largest in the country running from Houston, Texas, 1,600 miles to Linden, New Jersey. It is a common-carrier pipeline that is jointly owned by nine of the major oil companies. Each of the owners has a prorated share of the pipeline ownership which contributes to status quo relationships. At the terminal centers along the pipeline, different majors maintain their own storage and distribution facilities. Exchanges are arranged among majors along the line where one oil company has a more economical reach to a market than another.[13]

In contrast to the Colonial Pipeline is the Williams Brothers Pipeline. Since this pipeline is independently owned, major integrated oil companies, small integrated oil companies, and independent refineries can all tender their gasoline to the pipeline company at receiving points according to the pipeline specifications. The refineries arrange to take receipt of a like grade and quantity of product at various pipeline common-carrier terminals. The refineries pay a published tariff to the pipeline company for providing product at different distribution points. Price marketers have historically had adequate supply and have thrived in much of the ten-state area reached by the Williams Brothers Pipeline. The story is similar in certain other markets where transportation and terminal facilities have permitted price marketers to obtain an economical supply of unbranded gasoline.

Decline of Independent Refineries

THE SECOND principal reason for the reduction in the supply of unbranded gasoline has been the decline in the independent refiners. With crude-oil prices increasing approximately 150 percent from 1945 to 1957, as the major integrated oil companies shifted industry profits back into crude-oil production, a severe financial squeeze was exerted on much of the independent-refining segment of the industry. As a consequence, many independent refineries sold out to integrated oil companies with substantial crude-oil positions that did not look upon refining as an activity that had to carry itself, but more as a vehicle to insure an outlet for its profitable crude-oil operation.

Many of the remaining independent refineries are marginal operations. These refineries continue to operate because of their sunk investment in assets not because of the return to be earned from the activity. The continued operation of many of the independent refineries depends upon the subsidy they receive in the form of a bigger relative import quota which lowers their raw material cost to a more realistic level. In addition, a few independent refineries continue to operate in the north-central part of the country because of their access to low-cost Canadian crude oil. Thus, the continued existence of many of the remaining independent refineries precariously depends upon special considerations under the import quota system.

Effects of the U.S. Government's Clean-air Policy

THE MAINTENANCE of an adequate and economical supply of unbranded gasoline may become a critical problem to many price marketers in the early 1970's. The unbranded supply problem has suddenly become quite serious for many price marketers as a result of the U.S. government's clean-air campaign. The EPA has set standards requiring the emissions of carbon monoxide and hydrocarbons from light duty vehicles and engines manufactured during or after the model year 1975 to be reduced by at least 90 percentum from the emissions allowable under standards applicable to 1970 model cars.[14] In the process of moving toward a ban on leaded gasoline, the government has established conditions which could result in further suppression of price competition and lead to more industry concentration.

The problem of many of the price marketers is no longer simply the availability of gasoline, but of unleaded and low-leaded gasoline. Unless the price marketers are able to obtain these gasolines on some equitable basis their ability to compete will be severely impaired. The 1971-model cars are designed to operate on low-lead or no-lead gasoline as will probably be all new cars in future years. The development of low- or no-lead gasoline strongly favors the major integrated oil companies. With large industry profits locked securely away in crude-oil production and with huge associated cash flows, the major integrated oil companies are in the position to make the large investments required to change their refineries to produce the new formula gasoline. By early 1971 most of the giants of the industry were promoting and selling low- and no-lead gasoline.

The major oil companies with low level crude-oil self-sufficiencies, and particularly the independent refineries with little of their own crude oil, are hard pressed to finance the changes in their refining operations to produce low- and no-lead gasoline. As a result, they are slow in making the required modifications in their refineries; those with already marginally profitable operations will be forced to close as the sales of unleaded gasoline grow. Since such refining operations are a principal source of supply of unbranded gasoline, this avenue for a competitive and adequate supply of low- and no-lead gasoline for price marketers does not appear good. Furthermore, the major integrated oil companies that do supply some unbranded gasoline on a contract basis or that sell unbranded gasoline through independent agents have indicated in many cases that only when their own needs are satisfied will they furnish the new specification gasoline to price marketers. As the unleaded gasoline market grows, and the ban on leaded gasoline sales nears, a real squeeze may be exerted on independent price marketers unless some provision is made to insure them an economical and adequate supply of low-lead and unleaded gasoline on an equitable basis.

CONSEQUENCES OF TAKING PROFITS PRIMARILY AT THE CRUDE-OIL LEVEL

THE STRATEGY OF THE LARGE vertically integrated oil companies working in conjunction with government policy, which is discussed in Chapter 10, has been one of administering high and stable crude-oil prices and of taking the profits from the vertical system primarily at the crude-oil production level. Some of the consequences of this strategy have been:

1. high return on investment of independent crude-oil production companies;
2. a squeeze on independent refiners resulting in many sell-outs and a decline in the importance of independent refineries;
3. backward integration of vertically integrated companies to improve their crude-oil self-sufficiency position;
4. takeovers and mergers of vertically integrated oil companies with low crude-oil self-sufficiency that were caught in the squeeze between high crude-oil prices and low market prices; and
5. a reduction of economical sources of unbranded gasoline for price marketers and increasing dependency on major oil companies for supply.

Conclusion

Vertical integration in the petroleum industry has thus become a malignant force for harnessing monopoly power and for transmitting it to markets which would otherwise be workably competitive. So long as the present structural arrangement persists, workable competition in refining and marketing will always be frustrated. Many of the independents as well as integrated oil companies having a low degree of crude-oil self-sufficiency have been squeezed out of existence. Furthermore, the future for many of the remaining independents does not look bright. The petroleum industry is becoming more concentrated in the hands of those vertically integrated oil companies having strong crude-oil positions. If the petroleum industry is permitted to continue to administer artificially high crude-oil prices, competition in the industry will become still more limited.

10 / *Preferential Treatment of Crude Oil*

WE HAVE ALREADY SEEN that the absence of effective competition at one or more levels of an industry permits vertically integrated firms to concentrate industry profits in those activities, and in turn, to gain an important edge over certain nonintegrated competitors. In the oil business, it is the absence of effective competition in the production and purchase of crude oil that confers this strategic competitive advantage on the vertically integrated oil companies.

Competition had to be regulated for crude-oil production to become the profit haven of the petroleum industry following World War II, and the source of power and leverage. During the 1920's and early 1930's crude-oil prices fluctuated widely in response to new discoveries and economic conditions (see Figure 10-1). The first mechanisms for stabilizing crude-oil prices were developed in the early 1930's and were formalized with the passage of several state and federal laws regulating the production of crude oil. The laws enacted in the early and middle 1930's—state prorationing, the Connally Hot Oil Act, and the Interstate Oil Compact—were advanced as conservation laws. The oil conservation laws were supplemented in 1959 by the Mandatory Oil Import Quota System which was justified on the basis of national security.

OIL CONSERVATION LAWS

THE OIL-PRODUCTION CONTROL LAWS of the mid-1930's were passed as a result of the great economic waste associated with the reckless production of crude oil. Clearly the "law of capture," of oil belonging to the person pumping it, had resulted in costly over-drill-

FIGURE 10-1

CRUDE-OIL PRICE FLUCTUATIONS BEFORE AND AFTER BEGINNING OF PRORATIONING

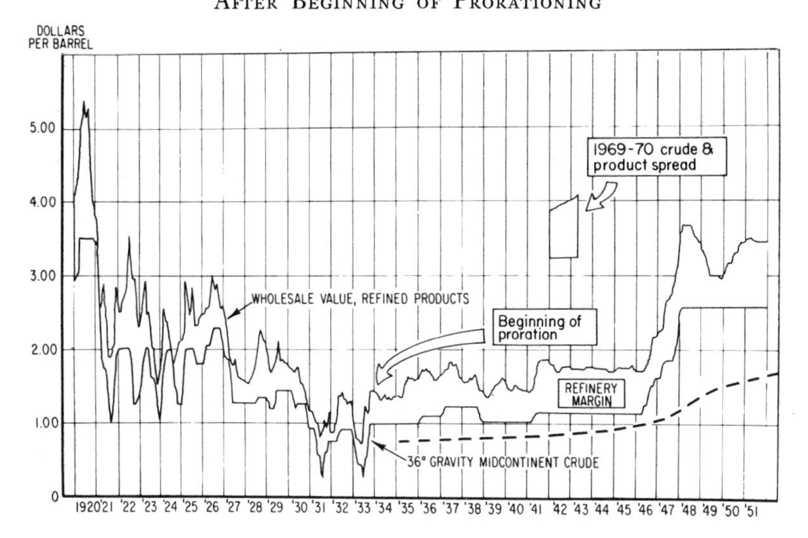

Source: Basic data from Ralph Cassady, Jr., Price Making and Price Behavior in the Petroleum Industry (*New Haven: Yale University Press, 1954*), *p. 136*

ing of oil fields and too rapid field production to bring about reasonable recovery of reserves of crude oil buried deep in the ground. Furthermore, the short-run over-production of crude oil had a very depressing effect on crude-oil prices. Undoubtedly the oil control laws did correct some of the excesses associated with the "law of capture." However, the oil control laws in the form they were passed went beyond conservation and tended to destroy the process by which competition determines price levels and left the pricing of crude oil to be administered by the integrated oil companies.

The oil conservation laws that were passed were principally formulated and lobbied for by the major crude-oil producers. Humble, the largest domestic crude-oil producer by the end of the 1930's, outlined a "conservation" program in 1927 that contained three essential features:

1. restriction of production to current demand as a measure of conservation;

2. regulation of production vested in state and not federal authority; and
3. creation of an interstate oil compact to allocate production among oil-producing states.

The general philosophy of the three features of Humble's conservation plan, along with many of the details, were incorporated in the set of laws passed five to seven years later to regulate the production of crude oil.[1]

State Control of Crude-oil Production

The regulation of crude-oil production is an activity primarily carried out by the individual states according to oil conservation and prorationing laws. The state prorationing laws which generally came into existence from 1932 to 1935 have two aspects: (1) maximum efficient rate (MER) prorationing and (2) market demand prorationing. MER prorationing recognizes the existence of pools of oil with multiple ownership. Each pool or reservoir of oil is considered a unit and given some maximum efficient rate of production which is further divided among the wells which have their own maximum efficient rating. The substitution of the maximum efficient rate prorationing for the "law of capture" has generally been recognized as a positive conservation development.[2]

Market demand prorationing is the procedure by which state oil-regulatory agencies restrict the quantity of oil produced during a month to approximately the amount demanded by crude-oil purchasers at the prevailing price. For example, the Texas Railroad Commission establishes the amount of crude oil that will be produced in Texas, basing its monthly allowables on estimates of monthly purchases by the major crude-oil buyers and regular forecasts of market demand by the Bureau of Mines.[3]

Many well-known oil economists recognize and accept the conservation reasons for MER prorationing. However, generally they are quite critical of market demand prorationing. The economists look upon market demand prorationing as the device used to substitute administered crude-oil prices for market determined crude-oil prices. They argue that as a result of the relatively fixed nature of demand for crude oil the major oil interests—independent crude-oil producers and the integrated oil companies with a high degree of crude-oil self-sufficiency—prefer higher prices which over time they have been able to obtain. Economists argue that as a result of market demand prorationing there are no price depressing short-run surpluses and that major oil interests can raise prices considerably above the level a free market would allow.[4]

The procedure by which market demand prorationing permits prices to be maintained above a free-market level is illustrated by the simple diagram shown in Figure 10-2. The short-run demand curve illus-

FIGURE 10-2

RESTRICTION OF SUPPLY UNDER MARKET DEMAND PRORATIONING PERMITTING PRICES TO BE ADMINISTERED ABOVE FREE MARKET DETERMINED LEVEL

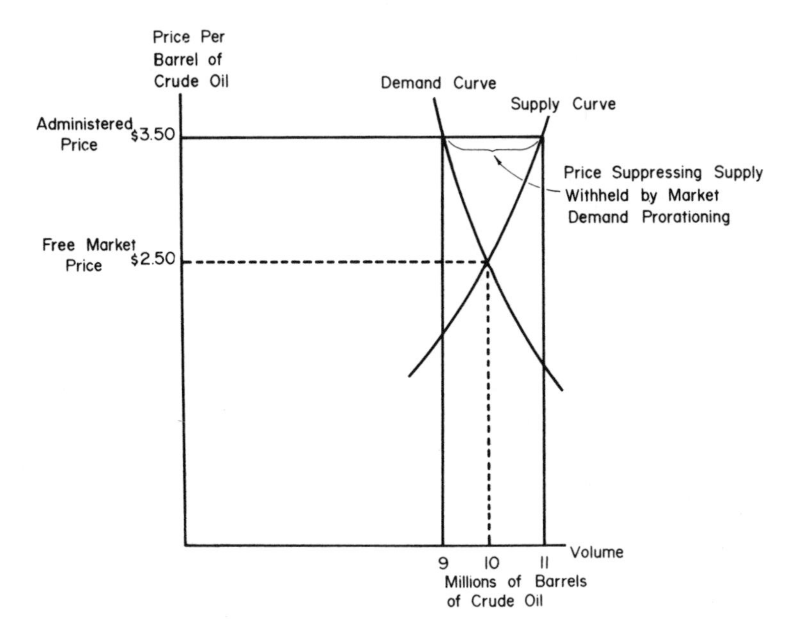

trates the quantity of oil that would be purchased at different prices; and similarly, the amount of oil that suppliers would be willing to sell at the different prices. In the absence of market demand prorationing the price of crude oil would fall in the example used from $3.45 per barrel to $2.60 per barrel. At the $3.45 price crude-oil producers would want to supply a quantity of 11 million barrels a day, but the demand would be 9 million barrels. The excess quantity supplied would eventually drive the free-market price to $2.60 per barrel where 100 million gallons would be supplied and purchased. However, since under demand prorationing supply is equated to demand at the price set by major oil interests, there is no excess supply to force prices to adjust to the free-market level.

Federal Laws and Agencies Support State Control of Crude Oil

OPERATING within their own powers, the states were unable to gain effective control of crude-oil production. The market demand prorationing states, those restricting production of crude oil, were confronted with the problem of (1) illegal production of crude oil beyond allowables that moved into interstate shipment and (2) coordination of production restrictions among states to insure that excess crude oil would not be produced. Therefore, the oil interests encouraged federal assistance in these areas to make state crude-oil regulation effective.

The first experience with market demand prorationing in the early 1930's clearly indicated that continued production of hot oil (that in excess of state allowables) would destroy the prorationing machinery. "The flow of 'hot oil' was declared to be comparable to the flow of bootleg liquor under prohibition."[5] In 1934 the federal government began to regulate the interstate shipment of hot oil in excess of production allowables of the states having market demand prorationing. On February 16, 1935, the Connally Hot Oil Act was passed by Congress which prohibited "the shipment or transportation in interstate commerce from any state of contraband oil produced in such state."[6]

The problem that remained after the passage of the Connally Hot Oil Act was that of some mechanism for forecasting the crude-oil demand and allocating production among the market demand prorationing states. This problem was solved when Congress authorized the Articles of the Interstate Compact to Conserve Oil and Gas as temporary legislation in August 1935. Since then there have been successive renewals of congressional consent first at two and then at four year intervals. One of the most important roles of the Compact is "voluntary interstate demand prorationing." The information gathering center which permits this coordination is the economic division of the Bureau of the Mines. This agency prepares at the request of the Compact monthly estimates of crude-oil demand which it allocates to the states. The annual forecast and production of the states are quite close. For example, a study covering the period 1946 through 1956 showed deviations of estimated and actual production of less than 1 percent as follows:[7]

8 years—Louisiana and Oklahoma
7 years—Texas
5 years—Kansas and New Mexico
4 years—Arkansas

Thus, in the case of Texas and Louisiana production of crude oil for eight of the eleven years was within 1 percent of the forecast. Clearly the Bureau of the Mines working with the Interstate Oil Compact plays an important role in an effective program for interstate market demand prorationing.

In summary, the system of state and federal oil laws has become a mechanism for permitting the administration of high noncompetitive crude-oil prices and for insulating crude-oil production from the normal competitive forces of the market. The artificial control of crude-oil production thus emerges as the crux of the perplexing problem posed by vertical integration in the petroleum industry. It means that government-abetted monopoly profits can be earned at this level, and that monopoly profits at this level can be used to affect competition adversely at the refining and marketing levels. It explains why profits are taken at the crude-oil level and why they are minimized at other levels.

Regulation of Oil Imports

During the 1920's and 1930's the United States was the oil capital of the world and the primary supply-balancing problem confronting the industry was domestic production. The unruly nature of domestic production was eventually solved by the passage and implementation of the so-called conservation laws. Following World War II foreign-produced and imported crude oil gradually added a new dimension to the crude-oil supply-balancing effort of the domestic oil industry. The supply problem posed by increasing crude-oil imports was finally solved by further government regulation of the competitive process of the oil industry.

A number of factors contributed to the increasing pressure of imported crude oil on the domestic oil industry following World War II. One of the fundamental pressures was the postwar discovery of huge low-cost oil reserves in Venezuela, the Middle East, and in North Africa.[8] While the United States consumed approximately 30 percent of the world's crude oil in 1969 its share of oil reserves declined to an estimated 10 percent of the total.[9] Another factor contributing to the growth of imports was the diminished control that the seven international oil giants had over free-world reserves outside the United States. From 1948 to 1969 their holdings of international free reserves fell from an estimated 90 percent to 70 percent. With this decline, their informal international prorating agreements—which regulated the international production of crude oil and made it possible to administer high worldwide crude-oil prices—began to collapse.

Another significant reason for the pressure of imports on the U.S. market was the doubling of U.S. crude-oil prices from 1945 to 1948 and the price increases of 1953 and 1957 which made the U.S. a high-price lush market for the low-cost foreign-produced crude oil. As a result the oil companies with low-cost foreign production (the international oil companies and the U.S. oil companies with some foreign oil such as Atlantic and Sun) had great economic incentives to substitute their low-cost imported crude-oil for the high-priced supplemental product they purchased from others to meet the crude-oil requirements of their refineries.[10]

Following World War II imports of crude oil and other petroleum products increased so that by 1948 they exceeded exports. Total imports rose from 377,200 barrels per day in 1946 to more than a million barrels a day by 1953, a quantity equal to 14 percent of U.S. production.[11] The imports continued to grow until 1958 when they had captured 18 percent of the U.S. market. Furthermore, the discovery of new, large low-cost foreign crude-oil reserves by new companies—Continental, Marathon, and Amerada—raised the likelihood of even greater imports in the near future.

The surge of imports of low-cost foreign crude-oil production into the artificially high priced U.S. market forced cut backs in U.S. production to keep supply and demand in balance. For example, during 1958 the Texas Railroad Commission, which regulates the number of days per month that controlled wells can produce, reduced production to eight days per month. Adding insult to injury was the actual landing of Middle East crude oil at Texas ports and the shipment of Venezuela crude oil up the Mississippi past idle Louisiana oil wells.[12] One of the strongest voices objecting to the increase of imports was the Independent Petroleum Association of America. The organization was founded in 1929 to represent the domestic oil interest in fighting the increasing imports of crude oil from Mexico. Early in 1953 the IPAA adopted a resolution urging Congress to restrict crude-oil and oil-product imports to 10 percent of the U.S. demand.[13]

In 1954 and 1957 efforts were made voluntarily to restrict the flow of low-cost foreign oil into the U.S. But the failure of the voluntary efforts and the prospect for still greater imports led President Eisenhower on March 10, 1959, to order mandatory quotas on imports of oil products. Imports of crude-oil gasoline and other finished petroleum products were limited to 12.2 percent of total domestic demand east of the Rocky Mountains while west of the Rockies imports were allowed as needed to supplement domestic production. The president invoked the import restrictions on the grounds that growing imports represented a threat to national security. Clearly both public and private interests were involved in the presidential decision but the private interests were especially strong.

The Mandatory Oil Import Restrictions Act of 1959 had the effect of isolating the U.S. market from the world market, which benefited the domestic crude-oil interests in two basic ways. First, the import quota system complemented the complex state prorationing system. It served to protect the crude-oil profit haven. With imports established as a fixed percentage of domestic demand, they could be systematically programmed into the production allocation formulas under state prorationing. Thus, the major domestic oil interests were able to maintain stable and artificially high domestic crude-oil prices since there were no disturbing short-run imbalances in crude-oil supply.

A second benefit of the import quota system to the domestic oil interests was that it effectively eliminated any ties between the U.S. crude-oil price and the free-world price. With the quantity of imports fixed, the domestic market would take no more than allowed under the quota regardless of how low the international price might be relative to the domestic price of crude oil. Since the invoking of the crude-oil import quota in 1959, domestic crude-oil prices had remained almost constant until increased by twenty cents per barrel in 1969 and twenty-five cents per barrel in 1970. In contrast, the international free-world price of crude oil declined approximately 40 percent from 1957 to 1969. The extent to which the import quota system has insulated the domestic crude-oil prices from the international price is reflected by the difference in the two prices delivered to a given point. For example, following the domestic crude-oil price increase of 1969, foreign crude oil landed in the major East Coast refining areas was $1.50 per barrel cheaper than domestic crude oil.[14]

Estimates of Costs of Import Program

THE walling off of the United States from the world market for crude oil has a huge price tag associated with it. M. A. Adelman, of the Massachusetts Institute of Technology, estimated the cost to be approximately $4 billion per year.[15] George P. Schultz, then Secretary of Labor and chairman of the Presidential Task Force established to study the import program, said in 1970 that if the quota were to be eliminated:

> American consumers would save about $5 billion annually now and over $8 billion annually by 1980. . . . On the basis of the report's figures the excessive cost to consumers for this decade would be in the neighborhood of $60–$70 billion. . . . I doubt whether the *cost of carrying such a subsidy for a single favored industry has ever been imposed on consumers by any Government, anytime, anywhere.* (Emphasis added.)

He went on to say:

> A majority of the task force found that the present oil import system does not reflect national security needs, present or future, and "is no longer acceptable." Its 12.2-percent limitation on imports into the bulk of the country is based on the mid-1950's level and has no current justification. . . . Besides costing consumers an estimated $5 billion each year (which we estimate will go to about $8.4 billion per year by 1980), the quotas have caused inefficiencies in the marketplace, have led to undue Government intervention, and are riddled with exceptions unrelated to the national security. Consumer prices for the whole country are, under the quota system, established largely as a result of limitations on oil production by one or two State bodies.

> To replace the present method and level of import restrictions, the report recommends phased-in adoption of a preferential tariff sys-

tem that would draw the bulk of future imports from secure Western Hemisphere sources. A ceiling would be placed on imports from the Eastern Hemisphere. These would not be allowed to exceed 10 percent of U.S. demand.

The tariff system would restore a measure of market competition to the domestic industry and get the Government, after a 3-to-5-year transition period, out of the unsatisfactory business of allocating highly valuable import rights among industry claimants. Tariff also would eliminate the rigid price structure maintained by the present import quotas. They would establish Federal rather than State control over this national security program with its important international implications.

An initial tariff level of $1.45 per barrel of crude oil (higher on products) would be established for imports from the Eastern Hemisphere. Further liberalization towards an equilibrium tariff level would be implemented (after further study) by the new management system proposed in the report. . . .

I personally am persuaded on the basis of presently available evidence that an equilibrium tariff objective of approximately $1 per barrel should be established now. It would not be reached for 3 or 5 years, depending on the transition period chosen, and the planning schedule could be altered by the management system if called for by counter-vailing evidence coming to light during that period. There is some uncertainty in our forward estimates now, but there always will be, and judgments still have to be made. My own judgment is that the national security—the only authorized basis for oil import restrictions—will be adequately protected by such a move. Believing that, I also believe it is fairer to the industry and to all affected interests if the objective is charted now.[16]

Decision to Preserve Import Program

THE recommendations of the Task Force (majority report) were consistent with the general position of the fourteen oil economists who testified during Senator Hart's antitrust and monopoly hearings in *Government Intervention in the Market Mechanism: The Petroleum Industry* in 1969. Not unexpectedly the recommendations of the Task Force were very much at odds with the position of the oil industry and their political constituents. Once the Task Force recommendations were leaked, President Nixon was bombarded with pressure from oil interests not to adopt them and to leave the import quota system unchanged. For example, Michael Haider, past chairman of the board of Standard Oil of New Jersey and retiring chairman of the board of the American Petroleum Institute, personally called on President Nixon and related his concern about the findings of the Task Force he had originally requested. Haider later reported to the API that he had had "a very good conversation" with the President and indicated that he was confident that the outcome would be favorable.[17] In addition, the governors of several of the oil-producing states, a number of senators, and many congressmen petitioned the president not to change the oil import system.[18]

Apparently these pressures were effective for when President Nixon released the oil-import study early in 1970, his statements indicated that he had decided against any immediate change. What the President did recommend was further study and discussion of the issues,[19] which was interpreted by some as a strategy of studying the issues to death. Later in October 1970 the president made it clear that there would be no immediate change in the import quota system when he announced that the system for oil imports would be continued into the "foreseeable future"—a move widely applauded by oil interests.[20] Oil's multimillion dollar campaign contribution to elect Nixon was paid back with huge dividends.

This decision to retain the costly oil import quota system leaves the domestic oil market isolated from the downward price pressure of the international oil market. As a result, the smoothly honed machinery of the domestic oil industry (resulting from state market demand prorationing, the federal Connally Hot Oil Act, and the federal Interstate Oil Compact) will continue to work to regulate the short-run supply of crude oil to demand. This means that the crude-oil price will be administered at its artificially high price level according to the joint interests of the independent producers of crude oil (the members of the IPAA) and the major integrated oil companies with strong crude-oil self-sufficiency positions. As a result industry profits will remain securely locked away in crude-oil production while the forward levels of industry activity will earn marginal or even negative returns.

Beneficial Tax Laws and High Crude-oil Prices

It was shown that an extremely important factor associated with the long-run success in the oil industry is for a major integrated oil company to produce a very high proportion of the crude oil needed by its refineries—to have a high degree of crude-oil self-sufficiency. A combination of state and federal laws has removed the determination of price from the normal competitive process and has made it possible for the major oil interests to administer high, stable, and noncompetitive crude-oil prices. In addition, crude-oil earnings benefit from a number of special tax laws which make it even more important for companies to own their own crude oil. The beneficial tax laws that will be discussed in this section include depletion, intangible tax write-offs, and foreign tax credits. These laws explain why many of the giant and wealthy oil companies pay very little federal income tax or none at all. For example, from 1962 to 1967, the very profitable Texaco Company paid federal income tax of less than 3 percent of its earnings while Atlantic paid no federal income tax at all on more than one-half billion dollars of earnings.

Percentage Depletion

One of the most valuable tax privileges enjoyed by the domestic oil producers is the depletion allowance. The theory upon which depletion is advanced is rather subtle. First, it is asserted that capital investment in an oil venture is quite a different thing from capital investment in a traditional business sense. For most business, capital investment is the cost associated with developing a producing asset. However, supposedly when oil is discovered, the oil man's capital is not the cost of finding oil, but rather what the "oil is worth in the ground." It follows that since oil is an irreplacable asset the owner should recover part of his capital with every barrel of production. The conclusion is then drawn that since income tax is a tax on income, it should be calculated after allowing for a deduction which is a return of capital—the depletion allowance. The leap of faith in this argument is in moving from a deduction against oil income of some portion of the actual capital investment (cost of incurring the asset) to that of a deduction against oil income of some portion of an imputed capital value (an amount several times the actual cost of the investment). What in essence this scheme amounts to is a tax-free recovery of the value of the oil deposit.

Enlarging the benefits of the depletion allowance. The depletion allowance first appeared as part of the tax law in 1913. From a minor tax loophole it has been nurtured by oil interests over the years to become a magnificent technique for reducing the income-tax liability on oil earnings by approximately 50 percent. In the beginning the depletion allowance permitted a deduction of up to 5 percent of gross income. However, it was limited to original cost or 1913 market value. In 1916 Congress legislated safeguards against tax deductions in excess of full cost and declared "that when the allowance authorized shall equal the capital originally invested, or, in the case of purchases made prior to March 1, 1913, the fair market value as of that date, no further allowance shall be made." [21]

As a war-time measure to encourage the search for new pools of oil, Congress passed the discovery depletion in 1918. For wells brought in in new fields, depletion was to be based on discovered value rather than being limited to actual capital investment. Thus beginning in 1918, depletion for wildcat wells was cut loose from actual invested capital and based upon a computed, engineered value. Within three years it was obvious to certain Treasury officials that discovery depletion had become a gigantic tax loophole. As administered, discovery depletion was not confined only to wildcat wells and new fields as had been the original intent. Furthermore, it was found that the inflated-value depletion write-offs against income frequently exceeded the income from a well. In an attempt to make discovery depletion more reasonable, Congress in 1921 limited the depletion write-off to a maximum of 100 percent of the profit from a well. Then in 1924 when Congress realized that many oil men were not paying any income tax on rather enormous earnings they limited the depletion write-offs to 50 percent of the profit from a well.

In 1926 significant changes were made in discovery depletion which broadened its coverage and standardized the practice. The most important change was the removal of the limitation of the special tax privilege to new oil discoveries so that all oil producers could now claim a depletion write-off. In addition, the concept of percentage depletion was adopted and the value of the write-off was fixed arbitrarily at 27.5

percent of the price of a barrel of crude oil, subject to the limitation that depletion not exceed 50 percent of profits.

The depletion law that evolved in 1926 remained fundamentally unchanged through 1968. As the corporate tax structure quadrupled from 13 percent in 1926 to 52 percent in recent years, the percentage depletion allowance became even more valuable to the producers of oil. During this period of forty-three years there were many attempts to "reform" the depletion allowance, but all the efforts failed. In part the preservation of the depletion allowance and other beneficial tax laws has been associated with oil men's spending some of the estimated $2 billion a year tax savings resulting from depletion to support the election of national officials sympathetic to oil's point of view.[22]

Example of how depletion works. A hypothetical example illustrates how percentage depletion works and how federal income tax is reduced. A profit and loss statement for the ABC Oil Company with six leases in three states is shown in Table 10-1. Gross income is $1,519,684 with expenses of $513,227, leaving a profit of $1,006,456 before deducting the depletion allowance and then calculating income tax. The tax law provides that oil income before tax may be reduced by the higher of percentage or cost depletion (notice recognition of difference in percentage and cost depletion). To determine the allowable depletion it is necessary to calculate 27.5 percent of gross income, 50 percent of net income, and cost depletion. (The revenue code of 1969 reduced the depletion allowance to a maximum of 22 percent.) From the example, the two productive Texas leases, nos. 1 and 3, take the full 27.5 percent depletion deduction ($101,793 and $73,509). The less productive and relatively more costly Texas lease, no. 2, cannot claim the full 27.5 percent ($40,435), and is limited to 50 percent of net income ($29,885). For the low producing New Mexico lease, no. 6, the cost depletion calculation ($15,975) is used since it is greater than the percentage depletion with the 50 percent net income limitation ($13,979).

For tax purposes the allowable depletion for the six oil leases is $374,573, which reduces taxable income from $1,006,457 to $631,884. The federal income tax is then calculated at a 52-percent rate to be $328,580 with profit after tax of $303,304. However, for accounting purposes and stockholder appraisal cost depletion (the actual book investment in asset) is used rather than the allowable value depletion

TABLE 10-1

TAX ACCOUNTING FOR THE ABC COMPANY

Leases	Gross Income	Total Deductions Before Depletion	Net Income Before Depletion	Depletion				Net Income After Depletion
				27½% of Gross Income	50% of Net Income	Cost Depletion	Allowable Depletion	
Texas:								
Lease #1	$ 370,157.00	60,416.70	309,740.30	101,793.18	154,870.15	56,617.45	101,793.18	207,947.12
Lease #2	147,035.00	87,264.49	59,770.51	40,434.63	29,885.26	27,995.46	29,885.26	29,885.25
Lease #3	267,305.00	47,424.73	219,880.27	73,508.88	109,940.14	17,203.45	73,508.88	146,371.30
Total Texas	784,497.00	195,105.92	589,391.08	215,736.69	294,695.55	101,816.36	205,187.32	384,203.76
Louisiana:								
Lease #4	109,652.00	141,609.62	(31,957.62)	30,154.30	(15,978.81)	7,862.28	7,862.28	(39,819.90)
Lease #5	529,265.00	108,201.09	421,063.91	145,547.88	210,531.96	—	145,547.88	275,516.03
Total Louisiana	638,917.00	249,810.71	389,106.29	175,702.18	194,553.15	7,862.28	153,410.16	235,696.13
New Mexico:								
Lease #6	96,270.00	68,310.82	27,959.18	26,474.25	13,979.59	15,975.35	15,975.85	11,983.83
Total New Mexico	96,270.00	68,310.82	27,959.18	26,474.25	13,970.50	15,975.35	15,975.85	11,988.83
Total leases	$1,519,684.00	513,227.45	1,006,456.55	417,913.12	503,228.29	125,653.99	374,572.83	631,883.72

Total allowable depletion: $374,572.88
Cost depletion: 125,653.99
Difference: $248,918.89

Source: Clark W. Breeding and A. Gordon Burton, Taxation of Oil and Gas Income *(New York: Prentice-Hall, 1954), pp. 192–93.*

(based upon the theory of "capital value," as previously explained). Thus instead of the inflated write-off of $374,573 for tax purposes, the actual write-off is $125,654. This means that the ABC Company recovered the difference of $248,919 without paying federal income tax. As a result the actual earnings reported to the stockholders are $552,223 rather than the $303,304 reported for tax purposes. The depletion allowance reduces the tax liability and greatly increases the after-tax earnings from crude oil. It is estimated that for the life of the average well that the inflated percentage depletion exceeds cost depletion (actual investment) by at least ten times. Depletion goes on and on over the life of the well which is quite different from the depreciation deduction allowed normal business operations or even the original law, which limited total depletion to the amount of the actual investment.

Incentives in depletion for high crude-oil prices. The mechanics of the depletion allowance and the preferential treatment of oil earnings give even more incentive to the vertically integrated companies with a high degree of crude-oil self-sufficiency to capture profits at the crude-oil level. In order to realize the maximum tax advantage crude-oil prices should be high enough so that net income before deducting depletion is 55 percent of gross income. This is shown in the example presented in Table 10-2. When oil is selling for $3 per barrel the allowable percentage depletion write-off is 27.5 percent. However, if the selling price were $2.50 per barrel the 50 percent of net income limitation would reduce the maximum allowable tax-free value depletion from 27.5 percent to 21 percent.

TABLE 10-2

DEPLETION INCENTIVE TO MAINTAIN HIGH CRUDE-OIL PRICE

	Maximum Tax Benefit		Partial Tax Advantage	
Price per barrel of crude oil	$3.00	100%	$2.50	100%
Less deductable expenses	1.35	45.0	1.35	54.0
Profit before depletion and income tax	1.65	55.0	1.15	46.0
Maximum depletion rate—27.5%	(.825)	(27.5)	(.688)	(27.5)
50 percent of net income limitation	(.825)	(27.5)	(.575)	(23.0)
Allowable depletion	.825	27.5	.575	23.0
Taxable income	.825	27.5	.575	23.0
50 percent corporate income tax	.412	13.75	.288	11.5
Profit after income tax	.412	13.75	.288	11.5

This simple example further demonstrates why vertically integrated companies having crude-oil self-sufficiency seek to maintain artificially high crude-oil prices. The inflated percentage depletion allowance reduces taxable income from oil earnings by approximately one-half so that the effective tax rate on crude-oil production is closer to 25 percent than the 50 percent charged earnings from other activities. Thus a dollar before tax earnings from crude oil (1.00 − .25 tax = .75) is worth approximately 50 percent more than a dollar of earnings from refining and marketing (1.00 − .50 tax = .50) on an after-tax basis. Therefore, companies that are reasonably self-sufficient in crude oil can increase the value of their pretax earnings by upwards to 50 percent if they shift them from refining and marketing to the crude-oil department. This is done by maintaining artificially high crude-oil prices in conjunction with the government, while holding refining and marketing margins at relatively lower levels.

Industry defense of depletion. Not unexpectedly, members of the oil fraternity and their trade associations have conducted over the years a major campaign in support of oil's special tax status. A long-standing argument in defense of percentage depletion is that investors need a special incentive to assume "special hazards" associated with producing crude oil.[23]

It is argued that in wildcat drilling—the exploratory drilling for new oil fields—frequently nine out of ten wells drilled are dry holes. The fact remains, however, that wildcat drilling amounted to only 15.5 percent of all wells drilled from 1961 to 1970. Approximately two-thirds of the total number of wells drilled were "developmental wells" in proven oil fields where roughly seven out of ten wells strike oil or gas. Overall, for the past twenty years, six out of ten wells drilled have been producers (see Table 10-3). Obviously the real risk in producing oil is wildcatting and not drilling developmental wells. If a special incentive is needed to encourage the discovery of new fields, presumably it should be restricted to wildcatters. This was the original justification for the incentive implied in percentage depletion. It is hard to justify this special tax benefit for all wells on the basis of the high drilling risk when six out of ten are producers.

In recent years the decline of drilling in the United States, which supposedly represents a threat for national security, has been used to defend the percentage depletion allowance. In fact many people have argued that it should be increased rather than reduced.[24] An interesting

TABLE 10-3

DRILLING ACTIVITY FROM 1951 TO 1970

	NEW WELLS DRILLED					
Year	Oil (No.)	Gas* (No.)	Dry (No.)	Service** (No.)	Total (No.)	Percent Dry
1951	23,453	3,030	16,653	1,380	44,516	37.4
1952	23,466	3,255	17,618	1,482	45,821	38.4
1953	25,762	3,806	18,449	1,262	49,279	37.4
1954	29,773	3,977	19,168	1,012	53,930	35.5
1955	31,567	3,613	20,742	760	56,682	36.6
1956	30,730	4,543	21,838	1,049	58,160	37.5
1957	28,612	4,626	20,983	1,409	55,024	38.1
1958	24,578	4,803	19,043	1,615	50,039	38.1
1959	25,800	5,029	19,265	1,670	51,764	37.2
1960	21,186	5,258	17,574	2,733	46,751	37.6
1961	21,101	5,664	17,106	3,091	46,962	36.4
1962	21,249	5,848	16,682	2,400	46,179	36.1
1963	20,288	4,751	16,347	2,267	43,653	37.4
1964	20,620	4,855	17,488	2,273	45,236	38.7
1965	18,761	4,724	16,025	1,922	41,432	38.7
1966	16,780	4,377	15,227	1,497	37,881	40.2
1967	15,329	3,659	13,246	1,584	33,818	39.2
1968	14,331	3,456	12,812	2,315	32,914	38.9
1969	14,368	4,083	13,736	1,866	34,053	40.3
1970	13,020	3,840	11,260	1,347	29,467	38.2

Source: The Independent Petroleum Association of America, "The Oil Producing Industry in Your State—1971 Edition," p. 104.

* Includes condensate wells since 1956.

** Includes stratigraphic and core tests.

inconsistency in the national defense argument follows from the fact that depletion is allowed on all oil production of all U.S. companies, either in the United States or abroad. A study done for the U.S. Treasury Department during the Johnson administration by the Consad Research Corporation found that percentage depletion was "a relatively inefficient method" of encouraging the search for new domestic oil reserves. The report indicated that "40 percent of depletion is paid for foreign production and nonoperating interests in domestic production." [25] This hardly squares with the national defense, high risk arguments.

Percentage depletion is fundamentally a subsidy to a particular class of businessmen—producers of crude oil. Furthermore, since companies vary greatly in their crude-oil self-sufficiency, the differential impact of the depletion subsidy has had a serious effect on competition in this industry. In general the huge vertically integrated oil companies with a high degree of crude-oil self-sufficiency, such as Texaco, Esso, and Gulf have thrived under such a tax structure. Also, the independent crude-oil producers have profited from the system of high administered crude-oil prices and special tax status. But, those vertically integrated companies with a low degree of crude-oil self-sufficiency, and the nonintegrated refiner-marketers, and marketing companies have been squeezed. Many with poor crude-oil positions were forced, as a consequence, to sell out and others are in serious financial trouble.

Intangible Tax Write-off

Another special tax privilege that oil producers enjoy is the option to expense, rather than to capitalize, most of the investment associated with the discovery of oil. This option allows the oil producers to quickly recover most of their investment in successful drilling operations. In contrast, the government requires that most ordinary businesses capitalize their investment in a producing asset and gradually write it off over a number of years.

The immediate recovery of approximately 80 percent of the investment employed in discovering oil is accomplished by the write-off against current income of the cost of dry holes and most of the intangible costs related to producing wells. For example, the cost of the search for oil and gas in 1964 (see Table 10-4) was $2,427,367,000 of

TABLE 10-4

Cost of Discovering Oil and Gas in 1964
(in thousand dollars)

Dry holes:	$ 853,808	35.2%
Producing wells:		
Intangible costs	1,131,228	46.6
Tangible costs	442,331	18.2
	$2,427,367	100.0%

Source: Petroleum Facts and Figures (*New York City: American Petroleum Institute, 1967*), *p. 30.*

which dry hole costs were 35.2 percent, intangible costs were 46.6, and tangible costs were 18.2 percent. Approximately 40 percent of all wells drilled were dry holes, and under the tax law the costs of the dry holes were immediately written off against current income.

While the write-off of dry holes is similar to the tax treatment of unsuccessful business ventures in general, the privilege of writing off intangible costs is quite another matter. The intangible costs amount to more than a billion dollars a year. This is approximately three-fourths of the cost of a producing well. Intangible costs, according to the tax regulation, are those costs having no salvage value, but which are "incident to and necessary for the drilling of wells and the preparation of wells for production of oil and gas." [26] Obviously, the intangible expenses are basic costs needed to bring in a producing well. In other industries, such costs are capitalized and depreciated over time since they are directly associated with the creation of a producing asset. As a result it is understandable why economists refer to the intangible tax write-off of oil as a "special privilege" in every sense.

Foreign Tax Credit

THE last of the "holy trinity" of beneficial oil tax laws is the foreign tax credit. The U.S. oil companies are permitted to make a direct offset against their U.S. tax liability for income tax payments made to foreign governments on oil earnings. This means of reducing U.S. income tax has been very beneficial during the past two decades to the major international oil companies—Standard Oil of New Jersey, Texaco, Gulf, Mobil, and Standard Oil of California—and in more recent years to a number of the principally domestic oil companies that have discovered some oil abroad. As a result of the foreign tax credit the major oil companies have been able to avoid most of their income tax liability that remained after it was first approximately halved by the percentage depletion allowance. From 1962 to 1968 the foreign tax credit wiped out approximately three-fourths of the remaining U.S. income-tax liability of the twenty largest integrated oil companies.

Prior to 1948 the payments made to foreign countries for the privilege of producing their oil were primarily on the basis of a royalty—the standard approach used to reimburse landowners. As new fields were discovered and production rose in the Middle East, the approximate 12.5 percent royalty on the low-cost and highly profitable production became unacceptable to the sheiks. They could calculate the vast profits the international oil companies were earning from their foreign production and they wanted a bigger share. In 1948 Aramco (owned by Standard Oil of New Jersey, Socal, Texaco, and Mobil) made a deal with King Ibn Saud of Saudi Arabia to split the profits 50-50, and the pattern quickly spread to the rest of the Middle East. Initially this cost the international oil companies very little for they were able to write off most of their foreign tax payment against their U.S. income-tax liability.[27] What in essence happened was that the U.S. government and the tax payer absorbed most of the cost of the profit-sharing deals made by the oil companies with the sheiks.

The approximate process by which the new plan worked is shown in Table 10-5. Assuming a $1.60 price per barrel of crude oil, operating

TABLE 10-5

DIFFERENCE IN PROFIT-SHARING ARRANGEMENTS AND ROYALTY PAYMENTS TO THE DOMESTICALLY BASED INTERNATIONAL OIL COMPANIES

	Before 1948		50-50 Profit Split		Increase in Royalty	
Price ber barrel of oil		$1.60		$1.60		$1.60
Operating costs	$.20		$.20		$.20	
12.5% royalty	.20		.20		.20	
Additional royalty	—		—		.60	
Total costs	—	.40	—	.40	—	1.00
Net income before tax		$1.20		$1.20		$ .60
Depletion (27.5%)		.44		.44		.30[1]
Income after depletion		$ .76		$ .76		$ .30
U.S. tax (50%)		.38		—		.15
Foreign tax (profit sharing)		—		.60		—
Net profit after tax		$ .38		$ .16		$ .15
Cash earnings:						
Depletion		$ .44		$ .44		$ .30
Earnings after tax		.38		.16		.15
Subtotal		.82		.60		.45
Applied residual foreign tax credit[2]		—		.22		—
Total		$ .82		$ .82		$ .45

1. Cannot claim full 27.5% depletion allowance because of 50 percent of net income limitation.

2. Sixty cents foreign tax, less $.38 offset against U.S. tax liability on oil, leaves a $.22 excess tax credit to be applied against other U.S. tax liabilities of companies from foreign operations.

costs and royalties totalling $.40, cash earnings before 1948 were approximately $.82 per barrel. Under the 50-50 profit sharing agreement, $.60 per barrel (one-half of $1.20) is paid to the sheiks in the form of a foreign tax. This tax payment is then employed as a direct offset against the U.S. tax liability of companies on foreign earnings. If the foreign tax is fully applied as an offset against U.S. tax liability on foreign earnings, then the 50-50 profit splitting deal costs the oil companies nothing. Fundamentally, what the 50-50 profiting plan amounted to was transferring the U.S. tax liability on foreign oil and other earnings to the foreign countries. In contrast, had the $.60 been paid to the sheiks in the form of higher royalties, the cash earnings would have been reduced by about 45 percent. Even if the excess foreign tax credit were not used, the 50-50 profit splitting plan would still be much more advantageous to the oil companies relative to the increased royalty payment plan.

Senator Edward Kennedy put into context the issue involved during the Senate Antitrust and Monopoly Committee Hearing.

> . . . The only question that I am interested in, is whether this should not really be considered as a royalty as far as our corporate income tax is concerned. On the one hand, if these payments are considered taxes, then the oil companies can credit them as dollar-for-dollar offsets on U.S. taxes owed. On the other hand, if they are royalties they can only be deducted. So, the question is not really whether a foreign country has the right to impose royalty payments. Obviously it can. But it is how we look at it, whether we are going to consider it as a royalty or whether we are going to consider it as a tax, and then if it is considered as a tax, it gets the dollar-for-dollar credit. If it were a royalty, it would be worth only 50 cents on the dollar as a deduction.[28]

Establishing a fictitious tax price. The profit-sharing plan worked very well with the sheiks for about ten years for much of what the oil companies relinquished was recovered by an offset against their U.S. tax liability. However, in the latter 1950's the foreign crude-oil price started to decline as the seven international oil companies began to lose their monopolistic control over Middle East oil. Initially price cutting took the form of small discounts from the posted price. Finally in February 1959 the oil companies reduced the posted price and did the same thing again in August 1960. The price cuts were very upsetting to the foreign countries for on their 50-50 profit-sharing program their take was reduced.[29]

The oil-exporting countries in September 1960 formed the Organization of Petroleum Exporting Countries (OPEC) to bargain collectively with the oil companies. From then on the oil companies were basically not permitted to reduce the posted price upon which the oil-exporting countries calculated their take. As a consequence, with falling real prices and fixed tax prices the 50-50 split became 70-30 splits. In the fall of 1970 Libya negotiated increases in the posted tax price first with the smaller oil companies and later with the large international oil companies. This spread to the other oil-producing countries. The net effect has been to further boost the profit-sharing percentages closer to an 80-20 split.

The foreign tax credit, the depletion allowed on foreign oil, and the intangible tax write-off has encouraged U.S. oil companies to develop foreign oil. Rather ironically the depletion allowance and intangible tax write-off had been justified on the ground of national defense and the need to encourage U.S. production. The foreign tax credit clearly works at cross-purposes to the encouragement of U.S. production.

Avoidance of Most of Their Income-tax Liability

THE net effect of the depletion allowance, the intangible tax write-off, and the foreign tax credit is to permit the integrated oil companies with a high degree of crude-oil self-sufficiency and substantial foreign operations to avoid most of their U.S. income-tax liability. For example during the period 1962 to 1968 the five U.S.-based international oil companies—Standard Oil of New Jersey, Texaco, Mobil, Gulf, and Standard Oil of California—earned $29.9 billion and paid federal income tax of $1.4 billion, a tax rate of only 4.7 percent. Other domestic oil companies with substantial foreign production such as Atlantic, Sun, and Marathon have similarly been able to offset most of their U.S. payment in one way or another. Overall the twenty largest integrated oil companies paid a federal income tax rate of only 7.7 percent on total earnings (see Table 10-6).

The administering of high crude-oil prices and the capturing of large industry profits in crude production which benefits from special tax treatment has placed the vertically integrated companies with a high degree of crude-oil self-sufficiency in a very favored competitive position relative to less well-balanced integrated companies and industry specialists who operate refineries and terminals and who are in marketing. Beside the better profit opportunities that exist in the production of crude oil relative to other industry activities, there are also strong cash-flow benefits derived from the special tax status of crude-oil earnings. The major integrated oil companies have historically "been able to provide nearly all [their] own capital requirements from . . . internal sources. . . ." [30] Being able to amass the necessary capital to keep pace with growing investment requirements without going to

TABLE 10-6

FEDERAL TAX PAYMENTS OF LARGEST REFINERS*
(millions of dollars)

		Net Income Before Tax	Federal Tax	%	Foreign, Some States' Tax	%	Profit After Tax
Std. (N.J.)	1962	1,271,903	8,000	0.6	423,000	33.0	840,903
	1963	1,584,469	69,000	4.3	496,000	31.0	1,019,469
	1964	1,628,555	29,000	1.7	549,000	33.0	1,050,555
	1965	1,679,675	82,000	4.9	562,000	33.0	1,035,675
	1966	1,830,944	116,000	6.3	624,000	34.0	1,090,944
	1967	2,061,000	166,000	8.1	700,000	34.0	1,195,028
	1968	2,313,587	233,999	10.1	802,907	34.0	1,276,681
	1969	2,069,697	265,789	12.8	756,269	36.5	1,047,939
	1970	2,474,748	268,273	10.8	896,938	36.2	1,309,537
Texaco	1962	546,371	13,000	2.3	51,700	9.0	481,671
	1963	615,768	10,250	1.6	58,850	12.0	545,668
	1964	660,761	5,500	0.8	77,900	11.0	577,361
	1965	681,613	10,000	1.4	80,700	11.8	590,913
	1966	808,991	32,500	4.0	104,700	12.9	671,791
	1967	892,428	17,500	1.9	124,400	13.9	750,528
	1968	1,006,246	23,800	2.3	162,800	16.1	819,646
	1969	952,854	7,250	0.7	175,800	18.4	769,804
	1970	1,137,666	73,250	6.4	242,400	21.3	822,016
Gulf	1962	488,351	19,389	3.9	128,871	26.0	340,091
	1963	540,065	30,870	5.7	137,842	25.0	371,353
	1964	607,343	52,443	8.6	159,782	26.0	395,118
	1965	655,727	53,559	8.1	174,935	26.0	427,233
	1966	813,868	90,008	11.0	219,098	26.9	504,762
	1967	955,968	74,142	7.8	303,539	31.8	578,287
	1968	977,321	8,005	0.8	342,997	35.1	626,319
	1969	992,005	4,264	0.4	377,183	38.0	610,558
	1970	990,197	11,892	1.2	427,939	43.2	550,366
Mobil	1962	379,339	8,300	2.1	128,700	33.0	242,339
	1963	437,352	23,000	5.2	142,500	32.0	271,852
	1964	464,660	27,700	5.9	142,800	30.0	294,160
	1965	508,016	33,900	6.6	154,000	30.0	320,116
	1966	555,412	23,200	4.4	176,100	31.7	356,112
	1967	594,593	26,900	4.5	182,300	30.7	385,393
	1968	673,739	22,000	3.3	223,500	33.2	428,239
	1969	728,815	41,800	5.7	252,500	34.6	434,515
	1970	873,744	95,600	10.9	295,644	33.8	482,500
Std. (Calif.)	1962	348,181	5,800	1.6	28,600	8.0	313,781
	1963	356,568	2,900	0.8	31,600	8.0	322,068
	1964	448,053	14,100	3.1	125,837	28.1	308,116
	1965	507,341	12,500	2.5	142,941	28.2	351,900
	1966	564,256	27,300	4.8	151,019	26.8	385,937
	1967	791,962	12,900	1.6	369,669	46.7	409,393
	1968	569,431	16,700	2.9	100,900	17.7	451,831
	1969	590,386	10,900	1.8	125,700	21.2	453,786
	1970	589,637	29,700	5.0	174,000	29.5	385,937
Std. (Ind.)	1962	168,843	3,105	1.8	3,381	2.0	162,420
	1963	208,022	22,182	10.6	2,748	1.0	183,092
	1964	204,817	8,486	4.1	1,480	0.7	194,851

TABLE 10-6 (continued)

		Net Income Before Tax	Federal Tax	%	Foreign, Some States' Tax	%	Profit After Tax
	1965	263,098	39,578	15.0	4,248	2.0	219,272
	1966	300,531	—	—	49,672	—	255,869
	1967	366,847	74,021	20.2	10,576	2.9	282,250
	1968	395,064	74,678	18.9	10,892	2.8	309,494
	1969	403,337	64,200	15.9	18,104	4.4	321,033
	1970	394,539	56,018	14.2	24,502	6.2	314,019
Shell	1962	173,555	7,200	4.1	8,680	5.0	157,675
	1963	211,575	19,100	9.0	12,623	5.0	179,852
	1964	213,575	2,800	1.3	12,583	5.0	198,190
	1965	274,507	26,600	9.6	13,876	5.0	234,031
	1966	313,085	46,100	14.7	11,785	3.7	255,200
	1967	342,022	44,940	13.1	12,233	3.6	284,849
	1968	387,767	63,378	16.3	12,298	3.2	312,091
	1969	308,451	5,464	1.7	11,836	3.8	291,151
	1970	274,681	34,285	12.4	3,191	1.1	237,205
Atlantic	1962	61,110	0	—	14,844	24.0	46,266
	1963	56,747	0	—	12,734	22.0	44,013
	1964	61,081	0	—	14,005	22.0	47,076
	1965	164,091	2,629	1.6	25,000	15.2	136,462
	1966	203,356	7,637	3.7	23,877	11.7	171,842
	1967	227,930	6,188	2.7	30,935	13.5	190,817
	1968	240,272	2,999	1.2	37,713	15.7	199,560
	1969	276,447	4,063	1.47	45,210	16.3	227,174
	1970	257,121	10,622	4.13	40,869	15.8	205,030
Phillips	1962	158,320	48,000	30.3	3,365	2.0	106,955
	1963	160,954	52,000	26.2	3,491	2.0	105,463
	1964	152,197	32,229	22.2	4,950	3.0	115,018
	1965	165,876	31,745	19.1	6,415	4.0	127,716
	1966	218,382	59,163	27.0	7,595	3.4	151,624
	1967	227,766	52,255	22.9	11,496	5.0	164,015
	1968	184,560	32,584	17.7	15,174	8.2	136,802
	1969	178,155	24,231	13.6	19,602	11.0	134,322
	1970	198,241	19,871	10.0	9,326	4.7	169,044
Conoco**	1962	73,477	1,065	1.4	3,335	5.0	69,077
	1963	99,665	9,143	9.2	3,157	3.0	87,365
	1964	204,184	235	0.1	103,840	50.9	100,109
	1965	201,914	4,670	2.3	101,093	50.1	96,151
	1966	290,924	11,669	4.0	122,339	42.1	156,916
	1967	280,584	7,649	2.7	123,973	44.2	148,962
	1968	290,357	9,721	3.3	130,594	45.0	150,042
	1969	248,868	3,394	1.3	88,406	35.5	157,068
	1970	301,115	119,262	6.4	121,388	40.3	160,465
Tenneco	1969	91,633	−13,299	—	2,941	3.2	101,991
	1970	182,082	24,273	13.3	2,891	1.5	157,809
Sun	1962	66,395	200cr	—	13,400	20.0	53,195
	1963	79,976	1,300	1.9	17,460	22.0	61,216
	1964	88,577	2,400	2.7	17,670	20.0	68,507
	1965	113,405	10,300	9.0	18,220	16.0	84,835
	1966	131,544	16,600	12.6	14,370	10.9	100,574

TABLE 10-6 (continued)

		Net Income Before Tax	Federal Tax	%	Foreign, Some States' Tax	%	Profit After Tax
	1967	146,946	24,700	16.8	13,670	9.3	108,576
	1968	227,790	44,290	19.4	19,070	8.4	164,430
	1969	226,052	48,207	21.3	25,585	11.3	152,260
	1970	223,086	27,054	12.1	56,957	25.5	139,075
Cities Service	1962	84,143	20,773	24.7	3,185	3.0	60,185
	1963	101,976	20,188	21.4	4,283	4.0	77,505
	1964	105,299	19,819	18.9	967	0.9	84,513
	1965	137,068	31,973	23.3	977	0.7	104,118
	1966	194,456	51,760	26.7	902	0.4	141,794
	1967	165,289	32,347	19.6	5,105	3.1	127,837
	1968	138,613	12,683	9.1	4,594	3.3	121,336
	1969	165,418	27,254	16.7	4,766	—	133,398
	1970	151,562	27,169	17.9	5,816	3.8	118,577
Union (Calif.)	1962	59,421	8,000	13.5	5,500	9.0	45,921
	1963	73,028	13,100	17.7	6,000	8.0	53,928
	1964	87,564	13,300	15.2	7,200	8.0	67,064
	1965	119,214	15,604	13.2	8,840	7.0	94,770
	1966	170,782	18,398	10.7	10,144	5.9	142,240
	1967	163,820	10,400	6.3	8,457	5.2	144,963
	1968	164,232	5,955	3.6	7,045	4.3	151,232
	1969	171,430	8,800	5.1	9,400	5.4	153,230
	1970	161,825	7,540	4.6	15,210	9.4	114,461
Amerada Hess	1969	133,875	2,406	1.8	46,813	34.9	84,655
	1970	183,905	6,648	3.6	63,247	34.3	114,010
Getty***	1967	132,762	3,687	2.8	10,909	8.2	118,166
	1968	112,798	6,712	6.0	7,836	6.9	98,250
	1969	140,426	11,400	8.1	9,758	—	119,268
	1970	159,144	34,909	21.9	13,089	8.2	111,146
Marathon****	1962	36,064	2,200cr	0.0	205	0.5	37,889
	1963	50,058	* * * *	0.0	933	2.0	49,125
	1964	63,220	* * * *	0.0	2,844	4.0	60,376
	1965	97,416	* * * *	0.0	37,345	38.0	60,071
	1966	130,927	2,400	1.8	59,700	45.9	68,826
	1967	138,520	3,700	2.7	60,962	44.0	73,858
	1968	155,335	4,350	2.8	67,659	43.6	83,326
	1969	170,657	3,250	1.9	77,929	45.6	89,478
	1970	153,783	8,200	5.3	61,000	39.6	84,583
Std. (Ohio)	1962	37,235	9,275	25.0	3,738	10.0	24,222
	1963	54,008	15,225	28.1	4,896	9.0	33,887
	1964	70,252	21,150	30.2	5,334	7.0	43,768
	1965	82,848	26,300	31.7	6,386	8.3	49,712
	1966	84,481	21,200	25.0	6,345	7.5	56,936
	1967	101,496	29,200	28.8	8,412	8.3	63,884
	1968	113,571	38,100	33.5	5,394	4.7	70,077
	1969	99,706	42,601	42.7	5,183	5.2	51,922
	1970	66,351	(6,918)	(10.4)	4,252	6.4	69,017
Ashland*****	1962	24,324	6,201	25.8	2,799	11.0	15,324
	1963	28,769	10,556	37.7	104	0.3	18,109
	1964	36,385	9,672	26.8	2,977	8.0	23,735

TABLE 10-6 (continued)

		Net Income Before Tax	Federal Tax	%	Foreign, Some States' Tax	%	Profit After Tax
	1965	50,594	15,500	30.6	2,440	5.0	31,594
	1966	69,324	20,830	30.0	5,570	8.0	42,924
	1967	72,212	23,718	32.8	3,952	5.5	44,542
	1968	79,115	26,251	33.2	4,524	5.7	48,340
	1969	83,102	27,480	33.0	3,280	3.9	52,342
	1970	84,326	27,260	32.3	4,723	5.6	52,343
Total	1962	4,198,331	169,492	4.0	838,954	19.9	3,194,770
	1963	4,921,577	304,985	6.2	951,255	19.3	3,663,037
	1964	5,322,329	233,241	4.4	1,251,442	23.5	3,838,846
	1965	5,940,312	407,621	6.8	1,360,470	22.9	4,172,221
	1966	6,985,171	577,436	8.2	1,609,663	23.0	4,798,072
	1967	7,626,420	628,581	8.2	1,945,888	25.5	5,051,641
	1968	8,131,063	623,458	7.6	1,983,326	24.3	5,524,279
	1969	8,031,315	589,454	7.3	2,056,265	25.6	5,385,595
	1970	8,857,753	774,908	8.7	2,463,382	27.8	5,597,740

* Reprinted with permission from *U.S. Oil Week* (October 25, 1971), pp. 5–7.

** Conoco's federal income tax figure includes a reduction due to benefits arising from consolidation. Foreign and state taxes include federal and state gasoline and oil excise taxes because the firm's financial statement gave no clear-cut breakdown.

*** Getty income for 1967 includes companies previously listed as Tidewater and Skelly.

**** Marathon Oil's 10K filing with the SEC doesn't reveal how much federal income tax it paid in years prior to 1966.

***** Fiscal year ends Sept. 30.

the capital markets has been another of the "wonders" associated with crude-oil production.

CONCLUSION

IN THE PETROLEUM INDUSTRY vertical integration is the crucial structural feature. Competitive behavior at each level is colored by the fact that the major competitors look at the effect of a competitive strategy on the profitability of the entire vertical system and not just at the profitabiliy of the level in question. A vertically integrated firm often makes competitive moves at a particular level which a nonintegrated competitor at that level finds impossible to follow or counter effectively. Marketing investments, for example, are made by the integrated firm primarily to move crude oil and only secondarily to earn a separately identifiable profit at the marketing level. For example, the major oil companies' practice of intensively blanketing a market with low-volume convenience stations only makes sense for an integrated firm which can increase its return at other levels. It would not make sense for a firm which operates only at the marketing level and has to earn a fair return from its investment in this activity.

Perhaps even more important, vertical integration in petroleum gives the integrated firm a strategic competitive advantage wholly unrelated to the efficiency with which it performs its operations or responds to consumer preference. The vertically integrated firm is often in a position to use its market power at one level to hamper its nonintegrated competitors at another. The state-abetted quasi-monopolistic position of the major firms at the crude-oil production level is extended by vertical integration to the potentially competitive refining and marketing levels. The net result is that the nonintegrated and partially integrated refiners and marketers are exposed to competitive pressures they find impossible to counter, regardless of their operating or marketing efficiency. As a consequence, vertical integration, coupled with the gross financial power and geographic diversity of the major petroleum companies has seriously impaired the competitive position of the independent refiners and marketers much beyond any difference in efficiency of size would warrant.

In the oil industry, the existence of the quasi monopoly at the crude-oil production level, coupled with preferential tax advantage, imbues competitive rivalry at the refining and marketing stages with a special aura of subsidy and unfair advantage that unduly and improperly constricts the competitive opportunities of independent firms, and all but destroys the principal sources of intertype competition and marketing innovation. At the very least it curbs their otherwise independent tendencies.

Given vertical integration, even such price competition between major integrated companies as sometimes accidently exists in product markets, works a differential harm on the nonintegrated refiners and marketers. Seldom does this price competition take the form of a change in interdepartmental billings or of an overt reduction in the posted prices charged nonintegrated operators in intermediate markets. More often it takes the form of a constricted margin in refining or a low-nominal profit rate in marketing. The inevitable consequence is a further squeeze on the margins of the nonintegrated refiner and marketer.[31] Therefore, competition between vertically integrated giants can only serve to destroy the smaller nonintegrated firm.

Above all else, however, the real threat which vertical integration in petroleum holds for workable competition is its tendency to narrow the availability of supplies and market access for those independent firms which promote their self-interest on a price basis and which are the principal sources of intertype competition. The strength and vigor of intermediate markets are the lifeblood of nonintegrated and partially integrated refiners and marketers. Viable intermediate markets are needed to assure customers an acceptable range of choices among alternative retail offers. Without real buyer alternatives it is impossible to determine whether the convenience and services showered on the motorist cost more than they are worth. How many gasoline service stations are too many? Only if the market is able to register free choice in these matters can private investment receive the guidance it requires.

In sum, competition in the oil industry is a victim of the industry's structure. So long as it persists workable competition at the retail level must remain frustrated.

11 / *Conclusions and Recommendations*

State of Competition

This study has focused on the vitality of competition in the marketing of gasoline to the public, and some of the conclusions reached indicate that there are many deep-seated problems in the industry. Marketing practices that appear to be designed to undermine free and informed consumer choice and that otherwise frustrate consumer sovereignty are widespread in the industry. Massive misallocations of resources—both of manpower and of capital—exist in the distribution of gasoline since the industry is not responsive to the pressures of the market place. Vertical integration, which is universal among the big firms in the industry, seems to be less of a response to technical and logistic imperatives than an attempt to transmit economic power from quasi-monopolistic crude-oil production to the potentially competitive marketing level. Prosperity is not necessarily associated with being progressive and efficient; rather it is due to market and financial power which tends to be derived from government policy that permits administration of artificially high crude-oil prices and which gives preferential tax treatment to crude-oil earnings. The recent rash of mergers involving many of the major oil companies and a host of the minor ones are resulting in structural changes in the industry that have undesirable short-run and long-run consequences for competition.

In short, this study reaches the conclusion that competition is not functioning as an effective regulator of the industry and that the principal companies in the industry have insulated themselves from the guiding influences of the market place. By whatever criteria one chooses to judge the marketing of gasoline—market structure, economic performance, or business conduct—the result is less than the public has a right to expect. Both prices and costs are higher than they would be if competition were working effectively. The entry of new firms and the emergence of new ideas and new methods are clearly restricted by the present structure of the industry. Predatory and exclusionary tactics to discipline innovators and to focus market and financial power on the smaller independents abound.

THREATS TO INTERTYPE COMPETITION

THERE IS REASON TO BE CONCERNED about the vitality of competition in the marketing of gasoline. Particularly important is the vitality of intertype competition between the major integrated firms with their fundamentally nonprice methods of marketing gasoline and the cadre of smaller independents with their price-discount marketing approach. It is the principal contention of this study that the marketing of gasoline in particular, and the performance of the industry in general, can be substantially improved if certain measures are taken to revive and otherwise to sustain the level of intertype competition provided by the smaller firms which have moved into the competitive void left by the major oil companies. An obvious corollary of this conclusion is that unless certain reforms are undertaken, this crucial form of competition will suffer further deterioration, and the marketing of gasoline will become even less responsive to changing consumer demand than it is at present.

The preservation and enlargement of intertype competition in gasoline marketing is especially important since gasoline retailing is about the only level of the industry still subjected to much price competition. Crude-oil prices are administered by the majors, pipelines are regulated by the ICC, and the refined product increasingly by-passes the open market and either moves through integrated distribution channels or is exchanged among the majors. Even at the retail level, the trend is away from price competition. Intratype competition among the major integrated oil companies while vigorous is primarily of a nonprice character. The emphasis is on the multiplication of modern well-located outlets, brand advertising, credit cards, games and stamps, esoteric additives, and image manipulation. The result is a homogeneous package of minimally differentiated products and services that are sold through a highly underutilized and inefficient distribution system at large cost to the driving public. The built-in costs and inefficiencies of the majors' approach to marketing make it necessary to sell gasoline at relatively high prices.

Majors' Commitment to Nonprice Competition

The major integrated oil companies cannot and will not, as a general rule, compete with one another by discounting prices. There are several reasons why they are exceedingly reluctant to try and improve their competitive position by tampering with prices. The giants of the oil industry are too big a factor in the market place to get away with much discounting of prices. A significant price reduction by a major oil company is almost sure to precipitate a retaliatory price cut by another major which would set off a price war where everyone loses. In another sense, the giants' operating styles and methods are too similar for anyone to have a significant cost advantage which would facilitate the discounting of prices. Furthermore, the major integrated oil companies gain from cooperating with one another in a variety of ways including the joint exploration for crude oil, the building of pipelines, and the exchange of product. It is inconsistent with this behavior for them to suddenly become very competitive in the marketing of gasoline on a price basis. For these reasons the competitive style of the major oil companies is primarily of a nonprice character.

Furthermore, price competition destroys the large operating margins and high prices needed for workable nonprice competition. Since the dominant method of selling gasoline has a low tolerance for price competition, the majors' marketing approach is designed to minimize the likelihood of a major-brand dealer voluntarily reducing price. This is in a large part accomplished by the major oil companies being supplier-landlords. The major integrated oil companies not only supply the gasoline, but also control most of the real estate and stations which they lease to so-called independent dealers. Through short-term leases, close supervision, and narrow dealer margins most of the major oil company dealers are discouraged from active price competition.

Role of the Independents

The independents operate in the market void created by the major oil companies' adherence to a fundamentally nonprice marketing approach. By operating relatively few, centrally located, high-volume, low-cost stations the independents have been able to sell gasoline on a discounted price basis. It is not because the independent price marketer is more public spirited or more inclined toward competition that he sells for less. It is because he can generate profitable volume this way.

The independents thus serve a competitive function in the gasoline business which is completely disproportionate to the share of market they command. They generally discount regular-grade gasoline by two to five cents per gallon with frequently a greater discount on premium-grade gasoline. The actual and potential gasoline volume that the price marketers drain away from the majors acts as a check on how far the majors can go with the nonprice method of marketing gasoline and how high they can set their prices. It is because the independents have to and are willing to reduce prices to compete that they play such an important role in this industry.

In one sense, the independent marketer does not have a price of his own. Rather, he discounts from the majors' price. Yet, this is not the complete picture. If it were, the independent's competitive behavior would not effectively serve to check attempts by the majors to increase their retail price. The independents would merely maintain their differential under the new higher price. Fortunately, the customer appeal of the price marketers' offer improves as the spread between their prices and those of the majors increase. Therefore, the price marketers gain by keeping their costs and prices low and by enlarging the differential as they have in recent years. It is this volume drain created by the price marketers that acts as a ceiling on how far the majors can go with their costly nonprice marketing methods and how high a level of prices they can establish for their gasoline. Presently one of the principal problems confronting the major oil companies is the need to check their increasing marketing costs. Both public and private statements by oil company executives recognize that major marketing costs are too high and have to be brought under control or more volume will be lost to the price marketers.[1]

Of equal importance to the price-regulating role of the independents is their role as marketing innovators. As a general rule, those firms which are the dominant competitors in any line of business—gasoline included—have a stake in the status quo. They frequently resist marketing innovation because if the innovation is a success others will have to copy them and their gains will be short-lived. On the other hand, the outsider has nothing to gain by maintaining the status quo and is motivated to introduce new marketing methods. This has been true in food retailing, drug retailing, and in the recent development of the self-service discount department store.

Intertype competition, thus, plays an important role in regulating the marketing of gasoline. Without the presence of a multitude of small aggressive price marketers, there is every reason to suppose that the performance of the industry would be much less acceptable than it presently is. They provide much needed assortment variety, i.e., they give the motorist a viable alternative to the homogeneous high-priced full-service offer presented by the majors.

Techniques Used to Thwart Intertype Competition

The problem is that this crucial intertype competition is threatened. It is being threatened in a large part because of the success of the independents in attracting business during the 1950's and in some areas in the 1960's. In those areas where the independents were making significant inroads into markets, the major oil companies reacted. One of the principal responses by the major oil companies to the encroachment on their share of the market was regionally concentrated price wars and prolonged periods of sales below reasonable costs. The techniques for executing this strategy have been price protection and zone pricing. The problem is not that the majors used price cuts to respond to the volume inroads of the price marketers. Rather the problem is that by resorting to price protection and zone pricing, the majors have been able to limit the scope of their price cuts and to focus the full force of their superior financial and economic power on selected independents to regulate competitive practices. Unfortunately, the result of this response has been destructive to intertype competition and a deemphasis of the normal regulating role of the market place.

The majors engaged the price marketers in the Mid-continent, the West Coast, and other pockets of strong price competition by selectively reducing prices through the technique of granting price protection. In many cases the independents were no match for the major oil companies who could draw upon their superior financial strength and functional and geographic diversity (for some of the majors, international diversity). The consequence of years of prolonged price wars in select markets and areas was that large numbers of independent refineries, terminal operations, and price marketers were forced to sell their operations.

There are other threats to the continued vitality of intertype competition. Important among them is the continuing decline of a competitive supply of unbranded gasoline. The independents must have access to an adequate supply of unbranded gasoline at prices which permit discounting if important intertype competition is to be preserved. The supply situation has worsened during the last decade and a half. The decline in the importance of the independent refiner and the rash of mergers of regional oil companies have eliminated several sources of supply of unbranded gasoline. The unbranded supply problem promises to become even worse as the ban on unleaded gasoline sales nears.

Price wars and inadequate supply pose real threats to the vitality of intertype competition in gasoline marketing. But of all the threats which bode ill for the industry vertical integration is the most significant and intransient. Unless structural reform is introduced, the long-run prospect for the independent price marketers is not bright. In a real sense, the other threats to intertype competition (i.e., price wars and supply) are merely symptoms of the structural problem confronting the industry. Yet, if one reads the history of business-government relations in this industry, structural reform (for example, divorcement) has little if any chance of gaining congressional or judicial support. Thus, while structural reform is recommended, it is seen as a long-run solution to the problem of preserving intertype competition. For this reason, most of our recommendations for improving the competitive performance of the gasoline industry deal with matters of business conduct.

RECOMMENDATIONS TO IMPROVE COMPETITIVE PERFORMANCE

THE RECOMMENDATIONS WHICH FOLLOW are considered central to the preservation of intertype competition with its role of naturally regulating competition in the gasoline industry.

Recommendation One

THE MAJOR INTEGRATED OIL COMPANIES SHOULD BE REQUIRED TO DIVEST THEMSELVES OF THEIR CRUDE-OIL HOLDINGS

WITHOUT the presence of horizontal market power at one level of an industry, vertical integration is not necessarily anticompetitive. In fact, an argument can be made that in the absence of horizontal market power at some level, vertical integration might actually foster competition and increase distribution efficiency. It is probably true that by itself the integration between refining and marketing in the petroleum industry has resulted in improved coordination in the logistics of gasoline marketing and has contributed to a general lowering of distribution costs. Unfortunately, when integration between refining and marketing is coupled with integration back into crude-oil production, the whole picture changes.

In their major study, de Chazeau and Kahn argued persuasively that the structure and practices of the petroleum industry beyond the crude-oil production level are "workably competitive." But they hasten to note that this otherwise benign integration becomes quite anticompetitive when coupled with crude-oil production. This backward integration allows the major firms to merge the monopolistic and competitive levels of the industry.[2] The result is that firms not so integrated are confronted with price squeezes and unfair competitive pressures they are incapable of resisting regardless of comparative efficiency.

In the petroleum industry vertical integration is not an inevitable response to the logic of increasing efficiency with the benefits shared by sellers and customers alike. Rather, it is a calculated strategy designed to exploit a state-abetted quasi monopoly in the production and sale of crude oil by carefully regulating downstream activities so as to preserve the benefits of the monopoly in crude oil and to maximize the integrated firm's overall profits. Without the major oil companies' integration from crude production through branded service stations, crude-oil prices and profits would be considerably lower since crude oil would have to be sold to independent buyers who would use their buying power to bargain for a fairer share of industry profits. They would seek to play the sellers off against each other to obtain lower raw material prices. Thus, vertical integration forward from crude tends to protect artificially high crude-oil profits from competitive erosion and makes it nearly essential for an oil company to have a high degree of crude-oil self-sufficiency to operate successfully.

Just as forward integration makes sense as a way of protecting artificially high crude-oil profits from competitive erosion, so does backward integration from refining and marketing into crude-oil production make sense as a way of escaping from the intolerable margin squeezes that competition with completely integrated companies inevitably imposes. The result of such competition is an irresistible pressure toward complete vertical integration in this business that is market-power centered rather than efficiency dominated. As the dominant companies in the oil industry approach more complete integration, it follows that they will seek to capture as much of the profits of the total system as possible at the protected crude-oil level. This practice is further encouraged by the preferential tax treatment accorded crude-oil profits under the federal income tax laws which permit oil companies to avoid most, if not all, income tax payments on crude-oil profits.

The logic of operating an integrated concern so as to maximize the profits earned at the noncompetitive, tax-favored crude-oil level has important consequences for the nonintegrated refiners and marketers and integrated refiner-marketers. Any competition that breaks out at the retail or refining levels will only result in constricted margins at these levels and will not result in a crude-oil price reduction. Furthermore, the nonintegrated rivals of the fully integrated firms will experience devastating price squeezes and will be subjected to unfair competitive pressures from rivals that not only don't have to make profits at each level, but don't even want to.

The integrated structure of the major firms thus frustrates potentially workable competition at the refining and marketing levels. Competition, which otherwise might be workable and constructive, becomes destructive for the nonintegrated firm when it occurs in a structural context which allows the major competitors to extend their market power at their crude-oil production level down into their refining and marketing operations. The result is unfair price competition and price squeezes that either force their nonintegrated competitors to integrate or to withdraw from the market.

Vertical integration allows firms with strong crude-oil positions to gain an unfair competitive advantage, which is completely unrelated to their comparative efficiencies, over their nonintegrated rivals. The tremendous pressure to move high-price crude oil directs major oil company competition along nonprice lines, resulting in the massive overbuilding of service stations and the development of costly marketing programs. This overindulgence in marketing with marginal or even negative returns makes competitive entry into the industry from retailing very difficult and frustrates the emergence of a countervailing buying power to oppose the monopolistic increase in crude-oil prices. The inevitable result is that in this industry the tendency of vertical integration is to narrow the availability of supplies and the access to the market for those independent marketers who sell primarily on the basis of price.

In summary, vertical integration poses two key problems—unfair competition and a drying up of the supply of economically priced unbranded gasoline. However, a reasonable structural remedy exists that would ease these problems. The structure of the industry beyond the crude-oil production level is workably competitive. Vertical integration between refining and marketing, without the market power which exists at the crude-oil production level, poses no serious anticompetitive consequences. It is the integration between crude-oil production and the refining and marketing levels that creates the critical problems in the industry. If refining and marketing were separated from crude-oil production, there would be no important power at either refining or marketing levels to use to gain unfair advantage at the other. In fact, even if vertical dissolution between refining and marketing were achieved, the structure probably wouldn't change much. Jobbers and retailers would still tie themselves to a single major-brand supplier. But this is not the case if the market power of crude-oil were divorced from the other activities. The structure would change dramatically. Nor would there be any great loss in economic efficiency if crude-oil production were separated from refining and marketing. Therefore it is proposed that the major oil companies be required to spin off their crude-oil holdings.

Vertical integration without market power at any level poses no real anticompetitive problem. Without the power at the crude-oil level, refining would have no base from which to subsidize marketing losses or vice versa. Competition from other firms at each level would serve to force each level to stand on its own feet. No longer would competition be unfairly subsidized by crude-oil profits. The competitive struggle between the majors and independents should be a fair struggle.

A number of important consequences would follow from the divorcement of the crude-oil market power from the rest of the industry. True, the market power at the crude-oil level would remain; divorcement wouldn't eliminate it. But, divorcement does something else nearly as good. It opposes it with countervailing power. Monopolistic sellers of crude oil would be opposed by strong independent buyers. These buyers would no longer have a stake in high crude-oil prices. Nor would they have an interest in curbing supply to maintain high crude-oil prices. Quite the contrary. They would now bargain aggressively to force crude-oil prices down, and they would probably make higher nominations so that the allowable production would be likely to produce a moderate surplus which would help reduce prices.

Without the crude-oil subsidy, normal market pressures would quickly cause the excessive investment in distribution to be reduced. Also, since crude-oil subsidies would no longer be available to cover refining and marketing losses, reasonable returns would have to be earned at each of these levels. Independent refiners and independent marketers would compete on equal terms with the integrated firms. Big firms would still have an advantage over small firms and regionally diversified firms would still be able to cover losses in one market with earnings from another. But these are not irresistible pressures and independent firms could survive. Also, with the price squeeze between the crude and the refined product removed, refining margins would rise and new independent capacity would emerge. This would serve to partially alleviate the unbranded supply problem.

While divorcement of crude oil is recommended, it is viewed as a long-run solution to the industry's problems. The courts have traditionally shown great reluctance in ordering divorcement as a remedy. If anything, the Congress is even less enthusiastic about such proposals. During the New Deal days an annual oil-industry divorcement bill was a regular ceremonial rite. Its prefunctory burial in committee was equally predictable. For this reason, several additional recommendations follow which deal with undesirable business conduct. These recommendations assume much greater importance without divorcement than they would if divorcement were a reasonable prospect.

Recommendation Two

PRICE PROTECTION AND ZONE PRICING AND OTHER FORMS OF DEALER SUBSIDIES SHOULD BE FORBIDDEN AS UNFAIR TRADE PRACTICES WHICH TEND TO LESSEN COMPETITION.

THE MAJOR integrated oil companies have extensively employed the technique of price protection to shelter and protect their nonprice method of marketing gasoline against price competition. What price protection amounts to is making selective price cuts in certain regions of the country, markets, or sections of markets. Very frequently the target of the selective price cuts are the price marketers. By selectively adding low prices to their costly nonprice methods of marketing in the areas where price competition is strong, the major oil companies are able to discipline and police price marketers and regulate their growth and the share of market they command.

The mechanism of price protection involves the selective reduction of prices to levels that, given the majors' cost structures, they would find generally impossible to maintain. Very frequently the level to which prices are lowered as a result of price protection is far below the level of reasonable costs of doing business. One of the principal reasons the major oil companies have for narrowly reducing prices is to take much of the profit out of genuine price marketing and to discourage competitors from employing the discount merchandising formula of selling gasoline on a high-volume, low-cost, low-price basis. Considered from various aspects, price protection is a technique of cross-market, or vertical, price subsidization with financial and economic power being marshalled to regulate the nature of competition in the market place.

Price protection was frequently used during the first half of the 1960's and once again in the latter part of the decade to finance regional, market, and localized price wars. The selective price wars figured importantly in the sellout of many significant independent refiners and thousands of stations selling gasoline on a volume price-discounted basis. They were generally taken over by vertically integrated companies that had profitable crude-oil operations and that were also geographically dispersed and thus less vulnerable to regional price wars. In addition, the regional price wars were a major contributing factor in the sellout of several large regionally concentrated integrated oil companies. Even these giants could not withstand the subsidized market prices with which they had to compete in their major markets.

Price protection has been a very effective device for combating price marketing and limiting its growth. It has been employed to subvert the normal forces of the market that naturally regulate competition. Specifically, it has been used by the major oil companies to sustain a top-heavy, inefficient, and enormously costly method for distributing gasoline. Without the artificial interjection of price subsidies into the market place, intertype competition would force many changes in the majors' costly methods of marketing gasoline and would bring about a general decline in the price of gasoline sold to the public.

Price protection schemes should be replaced by a system of pricing that cannot be arbitrarily manipulated to subvert the forces of the market place in regulating competition. It should be a simple and straightforward method of pricing which everyone understands, which treats similar types of customers fairly, and for which there are no secret rebates and special deals. One common method of pricing that would satisfy these criteria would be a pricing system based upon refinery prices for different categories of customers plus transportation costs. For example, each refinery might establish prices at which it would sell branded and unbranded gasoline to terminal operators, wholesale distributor accounts, and retail customers. Each type of customer would then add on the transportation cost of moving gasoline from the refinery to the petroleum terminals served by the refinery. Actually this method of pricing gasoline is similar to the procedure existing in the unbranded market for gasoline. Many unbranded gasoline customers start with the price of gasoline in the major refinery areas and add on transportation and other costs to the different markets to determine what they will pay for gasoline at different petroleum distribution centers.

The elimination of price protection and other arbitrary means of managing prices does not require the passage of new laws, merely the enforcement of existing statutes. There is abundant evidence that price protection is an unfair competitive practice and tends to lessen competition. In such cases the Federal Trade Commission has power to establish trade regulations prohibiting the practice. Even though the public gets a bargain in the short run, it must realize the frequent and severe price wars in certain markets that are financed by price protection are destructive to competition in general and intertype competition in particular and in the long run are contrary to the public interest. In short, price protection is a way of bringing the functionally and geographically diversified financial power of the major oil companies to bear on competitors in specific markets who would seek to compete by discounting prices. Price wars (both general and narrow) that are financed by price protection are one of the principal reasons that the marketing of gasoline is so rigid and inefficient. Without vigorous intertype competition the gasoline industry will retain its costly and generally ineffective system of marketing gasoline. Price protection tends to sap the vigor of this intertype competition.

Recommendation Three

ALL REFINERIES THAT ACCEPT THE SUBSTANTIAL BENEFITS UNDER THE OIL IMPORT PROGRAM SHOULD BE OBLIGED TO OFFER UNBRANDED GASOLINE FOR SALE ON AN OPEN COMMODITY MARKET IN AN AMOUNT EQUAL TO AT LEAST ONE-HALF OF THEIR CRUDE-OIL IMPORT ALLOCATION.

SINCE THE major refineries have been willing to participate in the import program in order to obtain substantial economic benefits, they should be obliged to use part of this benefit to improve competition. If refineries don't want to sell some of their gasoline in the open market, they need only give up their import quota. To the extent that they have been willing to accept government intervention in their operating affairs in order to qualify for imports, so should they be willing to accept government intervention to improve the vitality of competition. All sales should be made on a nondiscriminatory basis to the highest bidder. Price should be determined as in any other auction market. Price rigging and bid collusion should be prohibited as it is in other regulated commodity markets.

The continuation of price marketers as a viable marketing force is not only dependent upon eliminating the selective granting of price subsidies, but also upon continued availability of an adequate supply of unbranded gasoline at a reasonable price. Two principal developments in the industry during the last ten to fifteen years have given rise to serious concern about the adequacy of the supply of unbranded gasoline. First, the number of major integrated oil companies selling gasoline to unbranded accounts in significant quantity has decreased. Second, there has been a serious decline in the number of independent refiners who furnish unbranded gasoline. Should these two trends continue it is likely that the absence of an adequate economical supply of unbranded gasoline will severely cripple the ability of price marketers to keep price pressure on the industry.

One of the principal reasons for the reduction in the number of major oil companies that supply gasoline to price marketers has been the increased balance in their forward integration. Through both internal growth and merger many major oil companies are now able to sell most of their gasoline through their own controlled outlets. As balanced integration grows the amount of gasoline the major oil companies find necessary to make available to the unbranded market will continue to decline.

The cross-country and other market-expansion mergers of the integrated oil companies such as those involving Atlantic-Richfield, Sun, and Hess have also been a significant factor in the decline in the sources of supply of unbranded gasoline. Frequently those companies absorbed by mergers were important sources of unbranded gasoline. DX, which was merged with Sun, and Sinclair, which was principally divided between Atlantic-Richfield and Sohio-BP, were major sources of unbranded gasoline. Since these mergers, DX and Sinclair have been gradually withdrawing from the unbranded market. In general, this withdrawal from the unbranded market has taken the form of a nonrenewal of unbranded supply contracts, an increase in prices, or territorial restrictions on supply availability.

The reduction in availability of unbranded gasoline from the major oil companies is not only a matter of their balanced integration and broader market coverage. It is also a consequence of the basic policy of several of the major integrated oil companies not to sell to price marketers under any circumstances. It is the practice for majors such as Standard Oil of New Jersey, Texaco, Gulf, Mobil, Shell, and American to refuse to supply gasoline to the unbranded market. They will reduce refinery runs first. However, most of the majors freely exchange gasoline with one another since they have reasonable assurance that the gasoline will not be sold through independent channels at reduced prices. Exchanges, as opposed to open market transactions, further reduce the likelihood of surplus major-brand gasoline falling into the hands of price marketers.

Recommendation Four

THE U.S. GOVERNMENT, WHICH HAS MADE A NUMBER OF DECISIONS PLACING THE MAJOR INTEGRATED OIL COMPANIES INTO A PRIVILEGED CLASS, SHOULD TAKE THE STEPS NECESSARY TO INSURE THAT PRICE MARKETERS RECEIVE A FAIR SHARE OF LOW-LEADED AND UNLEADED GASOLINE AS THE BAN ON LEADED GASOLINE GROWS.

THE UNBRANDED supply problem facing the price marketers may become critical during the early 1970's. No longer is the problem simply one of the supply of leaded gasoline, but of the supply of low-leaded and unleaded gasoline. Beginning with the 1971-model cars, and on into the foreseeable future, all new cars will be built with engines able to operate on low-leaded and unleaded gasoline. As this happens leaded gasoline will gradually be phased out and there is a good possibility that a tax will be placed on leaded gasoline to speed the transition in use to low-leaded and unleaded gasoline. This governmental development establishes conditions which could lead to more integration and industry concentration as the restriction on sales of leaded gasoline grows. Unless the price marketers receive low-leaded and unleaded gasoline on an equitable basis, many will not survive the transitional period of the phasing out leaded gasoline.

The capital required to upgrade refineries so that they can produce high-octane gasoline without lead additives is substantial. The removal of lead from gasoline strongly favors the major integrated oil companies with their monopolistic profits and massive cash flows from the production of crude oil. Many of the remaining independent refineries are marginally profitable operations that will be unable to make the capital expenditure necessary to produce the new environmental grade gasoline and that will drop out of the refining business altogether.

With the new environmental grade gasoline contributing to a further reduction of independent refinery capacity, the price marketers will become more dependent on the major integrated oil companies for a supply of unbranded gasoline. If the price marketers are put on a "last serve basis," as some indications seem to suggest may be the situation, a supply squeeze could be exerted on the price marketers from which few will survive. Unless the price marketers receive an economical supply of low-leaded and unleaded gasoline during the transitional period many will be forced to sell out to the major oil companies, take major oil companies in as partners, or contractually integrate with oil companies by signing long-term purchase contracts. As a result of this combination of circumstances intertype competition could suffer a major setback.

Since it is not the intent of the U.S. government to injure the competitive forces of the market place in the process of removing lead from gasoline, provisions should be made to insure that there will be an equitable distribution of low-leaded and unleaded gasoline. One approach that the government might negotiate with the major oil companies and controlled refiners of gasoline is that they make available for purchase their low-leaded and unleaded gasoline to price marketers in the same proportion as their average sales of unbranded gasoline during the years 1968 to 1971. For example, if an oil company had sold on the average 30 percent of its gasoline to unbranded-gasoline customers before the development of low-leaded and unleaded gasoline, then it should make available 30 percent of the unleaded gasoline it produces for purchase by these accounts until the scarcity of low-leaded and unleaded gasoline passes.

In addition, the price of the low-leaded and unleaded gasoline must also be fair and reasonable or availability of unbranded gasoline will make little difference. In recent years, the wholesale cost of unbranded gasoline has generally been from one or two cents less than the wholesale price of branded gasoline and the same should hold true with unleaded gasoline.

Recommendation Five

THE JOINTLY OWNED FINISHED-PRODUCT PIPELINES SHOULD BE OPERATED AS TRUE COMMON CARRIERS.

THE TIGHTENING of the supply of unbranded gasoline is not only related to the change in the refining structure of the industry (the decline of independent refineries and the mergers of regional integrated oil companies), but also to the operation of the jointly owned finished-product pipelines. The major oil companies have been able to extend their quasi-monopolistic position in refining to the market place through control over the product pipelines. Through joint ownership and management of the pipelines, the partners in them are able to monopolize the only economical way to reach many inland markets. Those refineries that are not partners in the line most generally find that for all practical purposes they are denied economic access to the product pipelines. Non-partners face innumerable problems hooking into and getting off the product pipelines since the Interstate Commerce Act has no power to compel the building of facilities attached to the pipeline.

If jointly owned pipelines such as the Colonial and Plantation lines were operated as true common carriers, any refinery meeting technical specifications and reasonable minimum tender quantity would be able to use the pipeline on a fair and equitable basis. Each user would be charged a standard fee for handling its products and moving them to the carrier's output storage deposits along the line. In addition, the petroleum terminals along the pipeline might also be common facilities (as is the situation on the independent Williams Brothers Pipeline) instead of exclusive branded terminal outlets under the present arrangement. The way these interstate pipelines are currently operated supplements the control of the major integrated oil companies over markets and enhances their ability to regulate competition.

A rather dramatic difference exists in the nature of competition along the major oil company-controlled Plantation and Colonial Pipelines in contrast to the independently owned Williams Brothers common carrier pipeline which is used by the independent refiners and smaller integrated oil companies plus the major integrated companies. Along the major-owned and -controlled pipelines competition is in general mild in comparison to that which exists along the Williams Brothers line. If the Plantation and Colonial Pipelines were operated as true common carrier interstate pipelines, independent refiners and smaller integrated companies would have economical access to many markets now pretty much the exclusive domain of the larger cooperating members of the oil oligopoly. This would greatly increase the availability of unbranded gasoline supply and heighten competition along these pipelines.

Recommendation Six

THE MAJOR INTEGRATED OIL COMPANIES SHOULD BE PROHIBITED FROM ACQUIRING ANY INDEPENDENT REFINER, MARKETER, OR TERMINAL OPERATOR, OR ANY INTEGRATED OIL COMPANY.

WHILE SOME of the major integrated oil companies have nibbled away at the sources of supply of unbranded gasoline, others have been active in buying out unbranded terminal operators and marketing outlets for unbranded gasoline. For example, Esso and Gulf have purchased thousands of price marketer outlets and have converted them to their major-brand and nonprice method of doing business. The consequence of buying out the outlets for unbranded gasoline has been to reduce greatly the number of independent outlets in many parts of the country.

In those parts of the country where the buy-outs have seriously injured the role of independent marketing, the government should even consider breaking up the mergers and restoring the independents as a competitive force. For example, in Chicago the purchase of Oklahoma and Perfect Power by Humble and the Bulko stations by Gulf has removed from the market three of the major price marketers. As a result, today in many parts of Chicago drivers do not have the option of trading with a price marketer without driving considerable distance.

The buy-out of price marketers has been particularly strong on the West Coast. Thousands of price marketers have been purchased by major oil companies and converted to major brands. Price wars and price subsidies have weakened many of those that remain.

On the East Coast where sources of unbranded gasoline have always been scarce, significant independent terminal operations have been taken over by major integrated oil companies. Finally, large numbers of independent refineries have sold to integrated oil companies with strong crude-oil positions.

Also as noted in an earlier section, the mergers of regional integrated oil companies have serious implications for the availability of unbranded gasoline—especially if the merger produces a more balanced operation. Because of their importance as a source of unbranded gasoline, all mergers between integrated oil companies—especially the regional companies—should be forbidden.

Naturally this petroleum merger cessation recommendation doesn't stand by itself. The merger movement in the industry is in a large part a consequence of business conduct and industry structure. Therefore this recommendation is dependent upon some improvement in business conduct along the lines suggested by Recommendations Two through Five and even eventually on structured reforms as discussed in Recommendation One. In fact, unless some of the recommendations of this study are enacted, the merger movement can be expected to continue unabated with industry concentration increasing still further.

Final Statement

Some of these recommendations are bound to be controversial. Others have been repeated so often that they have become part of the traditional rhetoric of those who are concerned about what public policy toward the oil industry should be. Nevertheless, we think all of these recommendations are both vital and feasible. The probable social gains from adopting them far outweigh the social costs. Little in the way of systematic efficiency, scale economics, or entrepreneural motivation would be lost. Much in the way of increased intertype competition would be gained.

The primary purpose of the recommendations is the restoration of intertype competition as the principal regulator of business conduct in the marketing of gasoline. The specific restrictions on business conduct that are proposed are consistent with this purpose. Competition between large, financially powerful, vertically integrated, diversified organizations and small functionally specialized firms must not come to an unnatural end. Government intervention of some form is inevitable. Limited government intervention along the lines suggested will protect efficient independents against unfair and destructive practices and allow intertype competition to perform its role of keeping the industry responsive to changing consumer preference for new purchase combinations.

Appendix A

Letter Commenting on Characteristics of Early Self-service Gasoline

July 20, 1971

Dear Fred,

. . . YOU HAVE MISSED ENTIRELY THE ROMANCE OF THIS INDUSTRY. I do not think anyone will understand this phenomenon without a clear definition of the economic and human factors at work.

It should be pointed out that in those days service stations occupied corners with frontages as narrow as 40 feet; a 75-foot-frontage station was large; 2 to 4 pumps was maximum; underground tanks ran from as small as 500 gallons to 2500; bobtailed delivery trucks carried from 2500 to 3000 gallons; it took all day for a driver to deliver 2500 gallons, dropping 300–400 gallons at a time; to avoid running-over gasoline tanks, it was necessary to meter the gasoline out of the truck. In those days, service stations in California would not average over 5000 gallons per month.

The advent of serve-yourself was devastating. The Urich Oil Company chain of some 25 stations, of which I was Vice President, averaged 10,000 gallons per station. Our first serve-yourself reached 450,000 or so in three months, and our second, 570,000 gallons in three months. A 500,000-gallon station meant business failure to one hundred 5,000-gallon stations. The furor was unbelievable. Meetings of service station operators opposing us were held everywhere. One such meeting filled the Embassy Auditorium in Los Angeles. City Councils were deluged with requests for anti-serve-yourself laws. Eleven states and hundreds of towns outlawed our operation. By the end of the second year, or May 1949, I was spending half my time appearing before Planning Commissions, City Councils, Boards of Supervisors, Fire Commissions, and Chambers of Commerce. I received over forty death threats, and George R. "Frank" Urich received an equal number. In those days, he drove a truck and was fired upon on several occasions. I was run off the road by six men in a car who took off when I produced a 38 automatic. Our station help was beat up, creosol bottles were thrown against our buildings, and our neon signs were smashed with chains.

Appendix A

The world's first serve-yourself was opened by us in May 1947 on the northeast corner of Atlantic and Jilson in East Los Angeles. In those days of 40-foot service stations, all with islands parallel to the street, our first serve-yourself literally stopped traffic. Congestion was so great upon the lot that we installed a man on a small platform with a microphone to direct traffic. During the first three or four months we averaged one automobile wreck per day due to drivers' rubber-necking; on one day we had five automobile accidents fronting the station.

For a number of months thereafter, we employed a uniformed policeman who was stationed at the curb to direct traffic. Both sides of the street were parked solid for three blocks, primarily with oil company competitors. These men had counters to tally the traffic through our station. On the microphone we announced that we would give our gallonage out to anyone with a fill-up. Since the trucks were counted, we ran several empty trucks into the station each day, ostensibly dispensing gasoline into the tanks. The tanks were 10,000 gallons each (we now use 15,000 gallons) and were filled only when low enough to take a full truck and/or trailer load. No meters were used, cutting the delivery time down from an hour or more to 20–30 minutes.

During rush hours, girls on roller skates were used to collect from customers. The roller skates were taken off only during slow periods. Each girl carried a change-maker on her belt.

We gave away tickets with each purchase of gasoline. Actually, for legal reasons, we gave a ticket to anyone who entered the station and asked for it. These tickets were good for a drawing on a new automobile which was held every month, the program being continued for ten years.

In those early days of serve-yourself, oil was sold at the island but it was put into the motor by the driver himself, with the exception that oil sales were handled by the attendants for women drivers. All drivers were required to serve themselves for gasoline. The automatic shut-off nozzle was not available at the advent of serve-yourself and came in later.

In my appearances before City Councils, I read from a speech of Los Angeles County Fire Prevention Chief Thrapp wherein he stated that no special training was required by any service station employee; that boys of 16 to 18 years of age commonly worked in service stations; and that the customer spilled less gasoline than attendants since it was his money he was spending. I also introduced as evidence from 50 to 100 photographs of major oil company attendants smoking while delivering gasoline into the cars of customers.

By 1949 there were several hundred serve-yourself stations in California and the Serve Yourself and Multiple Pump Association was formed. The idea was conceived and the first meeting was called by Harry Rothchild, Sr.; hence, he can be called the father of the Serve Yourself Association although I served as the first President and remained on the Board for ten years.

The major oil companies did not believe the phenomenon would endure. Shortly after the opening of our first serve-yourself station, we offered the entire chain for sale, and Hancock Oil Company bought 25 conventional stations from us but refused to take the world's first serve-yourself. The sale money from Hancock served to build a number of other serve-yourselves and we were launched upon our way as the leaders of the independent gasoline industry on the West Coast.

For the first year following May 1947, we were regarded by the petroleum industry as some formidable breed of gangsters. Five years later, we were regarded as inspired marketers under the leadership of a retailing genius. In 1953 we sold 25 serve-yourself stations to Standard Oil Company of California, and Frank Urich and I were feted at a banquet given by Standard Oil Company at the Biltmore Hotel in Los Angeles where Mr. McClanahan, Standard Oil Vice President at that time said, "Urich Oil Company has taught the major oil companies their first new marketing lesson in forty years."

By the mid-50's the bulk of California independents were serve-yourself, patterned after our layout and design. The sheer number of serve-yourselves, and the competition between them, was so great that the bloom was off the rose, and we returned to windshield wiping and to dispensing gasoline when the customers so desired. Serve-yourself became a hybrid; by that I mean, the serve-yourself layout and image remained but the customer could serve himself or not; service would be given if he were willing to wait. We did not, and to this day, have not returned to lubricating cars and oil-changing. Even in our present operation, we sell gas, oil and additives only. We dispense oil in the motor but change no oil. We perform no other services and allow no tools on the premises. We do not sell any TBA (Tires, Batteries and Accessories).

During the first few months after the opening of the first serve-yourself, the station was circled during all daylight hours by aircraft photographing the station. We received written inquiries from major oil companies, as well as independents, from all over the United States and from foreign countries. Since, in the beginning, we stood alone against the wave of anti-self-service feeling, we encouraged competition and we mailed out over 2,000 sets of service station blueprints, free of charge, to any oil company official who asked for the same.

The serve-yourself, as it was conceived and created in 1947, served as the prototype of today's modern, large, so-called "multi-pump" station. The serve-yourself sounded the death knell for the 40, 50, and 60-foot station. As a consequence of this innovation, all stations have tended to grow larger and, consequently, fewer. Urich stations today operate throughout California, Arizona, Nevada and Oregon, and average 100,000 gallons. Station sites average 150′ x 150′ or larger. This is a far cry from the 5,000–10,000-gallon stations of 25 years ago. One driver of our truck and trailers today will deliver between 40,000 and 50,000 gallons of gasoline; the same driver 25 years ago would deliver 2,500. The large storage tanks were another innovation of my company.

You may be interested to note that in 40 years we have never had a service station fire.

Hugh Lacy
Senior Vice President

URICH OIL COMPANY

Appendix B

The Decline of Early Post–World War II Self-service Gasoline

By late 1948, less than two years after Urich opened his first station, self-service stations were estimated to hold over 5 percent of the gasoline business in the Los Angeles basin, one of the biggest gasoline markets in the world, and self-service operations were making sizable inroads in other markets where they had been introduced. The major oil companies were quite alarmed as they saw self-service operators draining away their gasoline volume and they took a number of steps to thwart the growth of the self-service concept.

On the local, state, and national levels effort was exerted to ban self-service operations. By the middle of 1949, at least twelve states had passed laws prohibiting them. When oil interests were unsuccessful at the state level they directed their efforts to seeking bans on self-service stations "city by city" and "county by county." For example, five cities in the state of Washington—Seattle, Vancouver, Longview, Olympia, and Everett—and Thurston County banned self-service stations by mid-1949.[1] On the national level The American Petroleum Institute, the primary association of major oil interests, passed a resolution in March 1949 calling for a ban on the sale of self-service gasoline.[2]

Harry J. Kennedy, then president of Continental Oil Company, had definite opinions about the wide-ranging efforts to kill self-service operations in the legal arena. He said that the efforts to outlaw self-service gasoline with state laws and local ordinances were misguided. "Self-service is a price problem, and no defender of free enterprise should ever consent to deal with price by legislation in peacetime." [3]

A second type of squeeze exerted on self-service operations involved the supply of gasoline. Since majors controlled most of the West Coast refining, many self-service operators were dependent upon them for their gasoline. This was true of two of the largest self-service operators—Urich and Golden Eagle. Urich had an agreement with Shell Oil to process its crude oil, and Golden Eagle purchased on contract from Standard Oil of California. Both of these contracts expired by the end of June 1949 and were not renewed. Urich started up its refinery that had been idle. However, Golden Eagle had no such option for it had closed and liquidated its refinery and turned over its crude-oil supplies to Standard Oil.[4] Supposedly in return Standard had indicated that it would never cut off Eagle's supply.

Standard offered to extend the contract for another year at an increase of 1.85¢ a gallon (approximately a 20 percent increase) and stipulated that Golden Eagle would not sell its gasoline for less than gasoline sold at Standard stations. Golden Eagle rejected the contract offered, claiming that it would put them out of business.[5]

A third problem facing self-service operations has been price wars. In the Los Angeles basin, majors and their dealers had generally maintained uniform prices during 1948 and 1949 at approximately 24.5¢ per gallon for regular gasoline and 27.5¢ per gallon for ethyl. Self-service operators used these pegged prices in announcing their savings of up to 5¢ per gallon of gasoline. With lower prices and the other attractive features, self-service operations readily grew. By the beginning of 1950 there were approximately 120 self-service stations in the Los Angeles basin as compared to 7,000 to 8,000 conventional stations. However, these few self-service stations were units whose combined volume represented about 5 percent of the market. Furthermore, many new self-service stations were being built and others were planned. The rapid public acceptance of self-service gasoline merchandising underscored the threat that this innovative approach posed to the major oil companies' method of selling gasoline.[6]

Toward the end of January 1950 Standard made its move by lowering prices in two of its California stations two to three cents per gallon—its break from uniform pricing.[7] During the next two weeks price cuts spread in locations where competition was particularly keen. Within a month the first phase of the price war was complete; it resulted in a cut in the retail price of about two cents per gallon. Estimates were that this involved about 90 percent of the 10,000 stations in southern California.[8]

Major dealers initially fought the price reductions, for most of the cut originally came out of their retail margins which they needed to cover expenses. A dealer association was formed in Los Angeles to unite dealers against price reductions that were not associated with a lowering of the wholesale price of gasoline. The battle was unsuccessful, for the reduction in price by one major station acted as a catalyst and set off another until the entire market collapsed.[9]

Prices at major stations in Los Angeles sank from a preprice-war level of 24.5¢ for regular and 27.5¢ for ethyl to 22.1¢ (down 2.4¢) for regular and 24.1¢ (down 3.4¢) for ethyl. This was similar to the price war that occurred throughout other parts of California and in other areas where self-service operations were making significant inroads. In Washington, Arizona, Wisconsin, Texas, Virginia, and New Jersey price wars raged as majors tried to corral self-service operations before they got too far out of hand. The length of the price wars varied from several months to more than a year. For example, prices were off 4.0¢ per gallon for approximately eight months in San Francisco.[10] In Los Angeles the price war that began in February 1950 continued with but few interruptions through January 1951 when the major stations hauled down their price signs in most areas.

Initially, dealers had cut prices out of their normal five-cent margin and were operating on a two- to three-cent margin. As the price war continued major suppliers found it necessary to give financial assistance to their dealers to keep them in business. However, a peculiarity of the Los Angeles price war was that the majors did not reduce the wholesale price (dealer cost). Instead, they allowed certain leasees price assistance when they lowered their prices to meet self-service competition. The aid of the majors to their dealers took the form of rental concessions, rebates, and competitive allowances. When the majors withdrew these special assistance plans, prices moved back toward normal.[11]

The prolonged price wars spelled the end of self-service gasoline on the West Coast. When self-service operators were competing with the dealers of the major oil companies, they had the upper hand and a revolution in gasoline distribution was in the making. However, when the major oil companies entered the picture the self-service operators proved to be poor adversaries.[12] Since the oil companies were integrated they did not primarily rely on marketing for their profits as did the self-service operators, and the major oil companies also gained financial strength from having widespread operations.

The major oil companies' financed price war brought economic pressure to bear on the self-service operators. As the major dealers lowered their price, the four- to five-cent price edge that many self-service operators originally had was narrowed to one to two cents, and at times self-service gasoline was even more expensive than major-brand gasoline.[13] In addition, competitive pressures forced some self-service operators to reduce their operating margin.

The reduction of the price edge rapidly drained off the high-volume business of self-service stations. Two hundred thousand gallons of gasoline per month in the preprice-war period was considered a good average and stations doing less than 100,000 gallons were disappointing. After only a few months of the price war 100,000 gallons per month looked good. For example, at the end of the price war the pioneer in self-service, Urich, had twenty stations doing approximately the same volume as its eight stations were before the price war.[14]

With lower volumes and operating margins, profits were hard to come by and red figures were common. It was this squeeze that finally forced most self-service operators to capitulate. Since the majors would not allow self-servers to operate on their original formula—a big price spread, large station volume, and a reasonable profit margin—new strategies had to be developed. Some of the self-service operators decided if they couldn't beat them, they would join them, and began to sell major-brand gasoline. Others increased their prices as the majors wanted and began to give service to their customers.[15] Urich Oil company made a number of changes which were similar to what others did. Urich reduced its six- and seven-island stations to four islands, for the extra islands were not needed and made its operation less efficient. In addition, Urich moved to a semiservice plan. Typically, during busy periods attendants would pump gasoline and collect money. During slack business periods attendants would give as much service as the customer wanted.[16] In addition, Urich commenced to sell tires, batteries, and accessories.

The final chapter in this early era of self-service operations was the sellout or the conversion of one-time self-servers to major oil companies. In November 1953, Urich Oil sold twenty-five of its thirty-one stations to Standard Oil of California.[17] Standard in turn converted these stations to the major style of doing business. The following year Golden Eagle, another of the early self-service operations, followed a course similar to Urich's. Eagle reached an agreement with Union Oil to convert 100 of its better stations to the Union 76 brand. Since Union "fair traded" its gasoline, the price of gasoline for the 100 ex-Eagle stations was increased to the level prescribed by Union Oil.[18] These two conversions to major operations materially reduced the number of independents selling to the public on a low-price volume basis.

The postwar boom of self-service stations was over by the early 1950's in most parts of the country. This era in the history of gasoline marketing demonstrated the possibilities of high-volume sales through a few outlets having relatively low cost and deep-cut prices. However, it also showed rather clearly that regardless of how efficient and economical the self-service method of selling gasoline, the major oil companies would not sit back forever and watch their volume be drained away to mass merchandisers of gasoline if there was something they could do about it. When the majors' losses became too great—the tipping point reached—they would find ways of lowering prices to certain of their dealers so that they could compete on a price basis with the mass merchandisers. The self-service operators then had basically two choices—lower their prices and try to maintain their volume, but give away their operating margin; or hold their prices and gross margin, but lose their volume. The consequences of either of these choices was a severe reduction in gross profit which would likely result in an unprofitable operation. The conflict was resolved by capitulation of the financially weaker and vulnerable operation. On the West Coast the losers of the battle were the self-service operators. The self-servers did not have the widespread market coverage of the major oil companies nor the multiple levels of industry activity from which they could draw financial support to sustain their marketing losses.

NOTES

1. "Self-Service Advocates Seek to Put Station Ban Controversy on the Election Ballot," *National Petroleum News* (July 1949), p. 20.

2. "API Board Hits Self-Service," *National Petroleum News* (October 26, 1949), p. 23.

3. "API Board Hits Self-Service," p. 23.

4. "Gasoline Contracts to 2 Suppliers of Self-Service in California Ended," *National Petroleum News* (July 6, 1949), p. 19.

5. Frank Breese, "Eagle-Standard Return Engagement Assured," *National Petroleum News* (August 3, 1949), p. 11; and "Eagle-Standard Oil California Conspiracy Case Hinges on 'Concert of Action,'" *National Petroleum News* (August 3, 1949), p. 15.

6. Frank Breese, "Self-Service Battle Each Other as Crowding Heats Competition," *National Petroleum News* (March 1950), p. 13.

7. "Standard Station Price Slash May Cut Self-Service Gallonage," *National Petroleum News* (February 1, 1950), p. 13.

8. "Dealers Fight Losing Battle to Hold Gasoline Prices Up," *National Petroleum News* (March 8, 1950), p. 11; and "Independent Dealers Rap Major Suppliers in Gas Price War," *National Petroleum News* (March 15, 1949), p. 15.

9. "Dealers Fight Losing Battle to Hold Gasoline Prices Up," p. 11.

10. Frank Breese, "Gasoline Prices Firmer in Many Areas, but Continue Soft in Los Angeles," *National Petroleum News* (September 6, 1950), p. 26.

11. Frank Breese, "Fading of Bitter Gasoline Price War Makes Dealer Rebates New Headaches," *National Petroleum News* (August 22, 1951), p. 26.

12. "Fading of Better Gasoline Price War . . . ," p. 26.

13. Frank Breese, "Multi-Pump Station Competition Forces Most Self-Serves to Give Some Service," *National Petroleum News* (September 5, 1951), p. 36.

14. "Self-Service Battle Each Other As Crowding Heats Competition," p. 13.

15. Frank Breese, "Self-Serves Keeps Customers Working Despite Mass Swing Back to Service," *National Petroleum News* (July 25, 1951), p. 28.

16. "Multi-Pump Stations Competition Forces Most Self-Serves to Give Some Service," p. 36.

17. "Urich Sells 25 of 31 Multi-Pumps," *National Petroleum News* (November 4, 1953), p. 24.

18. Frank Breese, "New Marketing Venture," *National Petroleum News* (March 24, 1954), p. 44.

Appendix C

Zone Pricing Data for Washington, D.C.

Competition along the four major arteries leading away from the center of Washington, D.C. illustrates the practice of zone pricing. Stations located along each of these arteries have been grouped for study in Table C-1. The first group of thirty-seven stations is located on Rhode Island Avenue running from Scott Circle, a distance of a few blocks from the White House, to the Circumferential Highway—Interstate 495. At the southern end of Rhode Island Avenue are three price marketers (a big canopied Martin station, a large Hess station, and an old White Seal station). The Martin and Hess stations are big-volume operations. The price marketers sell regular for 29.9¢ and 30.9¢ and premium for 33.9¢ and 34.9¢ (Martin also gives stamps). The prevailing major-brand service station price through greater Washington, D.C. is indicated by the darkened lines bracketing regular at 35.9¢ and premium at 39.9¢ and 40.9¢ shown in Table C-1. However, in the vicinity of the volume price marketers two of Esso's stations sell gasoline at 3.0¢ below normal and two sell it at 2.0¢ below normal; three of the four stations post their prices. Texaco has one station selling gasoline 2.0¢ below normal and posting its prices with stamp give-aways. Esso and Texaco have reduced prices by 2.0¢ to 3.0¢ to challenge the price marketers while the remainder of the competitors hold prices at normal (Gulf and Sun), or above normal (American) levels. There are no price marketers in the next group of eight stations and all but one of the stations sells gasoline at the normal major-brand price; and none post their prices which is similar to the pricing pattern of major-brand stations throughout Washington, D.C. where there are no price marketers.

On Georgia Avenue, which moves traffic north out of the city the same pattern can be observed. Major-brand prices in the group of stations near the two price marketers (Scot and Save More) have been reduced from 1.0¢ to 3.0¢ per gallon and approximately one-half of the majors post their prices.

The prices on Connecticut Avenue, where there are no price marketers in the three groups of major-brand stations, provide a contrast to the condition on Rhode Island and Georgia Avenues. Only two of the twenty-four stations are below major-brand normal and four are above it. Only one station posts its prices and the frequency of giving stamps among all stations is less. Furthermore, a majority of the stations have increased the regular premium differential from 4.0¢ (35.9¢-39.9¢ to 35.9¢-40.9¢) to 5.0¢ per gallon.

TABLE C-1

Cluster Analysis Of Competition In Washington, D.C.

Street	No. Stations	28.9¢ / 32.9¢ / 4¢	29.9¢ / 33.9¢ / 4¢	30.9¢ / 33.9¢ / 3¢	— / 34.9¢ / 4¢	31.9¢ / 35.9¢ / 4¢	32.9¢ / 36.9¢ / 4¢	33.9¢ / 37.9¢ / 4¢	34.9¢ / 38.9¢ / 4¢	— / 39.9¢ / 5¢	35.9¢ / 36.9¢ / 4¢	— / 40.9¢ / 5¢	36.9¢ / 40.9¢ / 4¢	— / 41.9¢ / 5¢	37.9¢ / 41.9¢ / 4¢
Regular		28.9¢	29.9¢	30.9¢	—	31.9¢	32.9¢	33.9¢	34.9¢	—	35.9¢	—	36.9¢	—	37.9¢
Premium		32.9¢	33.9¢	33.9¢	34.9¢	35.9¢	36.9¢	37.9¢	38.9¢	39.9¢	36.9¢	40.9¢	40.9¢	41.9¢	41.9¢
Differential		4¢	4¢	3¢	4¢	4¢	4¢	4¢	4¢	5¢	4¢	5¢	4¢	5¢	4¢
Rhode Island (north)	21 (3–18)		PB(5)		HESS		E E	E E			G1 G3			AM	
				MARTIN				T(5)	E	SI T SU3	SU3 SU3	AM(6) AM AM			AM
"	8 (0–8)							CI(5)			E SU4		G(3)3		
											E SH SU3 SH				
Georgia (north)	13 (0–13)										GU(3)2 E	AM AM			
									E	T GU4	SU3 CI E SU3 E SH				
	15 (2–13)					SCOT PB(5)	T	E5	SU2 E E SU2	AM	G3	AM AM SH			
										AM	E				
	9 (0–9)								T		E SU3 E SU3	AM			
											SI T		SH		

TABLE C-1 (continued)

Cluster Analysis Of Competition In Washington, D.C.

Street	No. Stations	Code	28.9¢	29.9¢	30.9¢	—	31.9¢	32.9¢	33.9¢	34.9¢	—	35.9¢	—	36.9¢	—	37.9¢
Premium			32.9¢	33.9¢	33.9¢	34.9¢	35.9¢	36.9¢	37.9¢	38.9¢	39.9¢	36.9¢	40.9¢	40.9¢	41.9¢	41.9¢
Differential			4¢	4¢	3¢	4¢	4¢	4¢	4¢	4¢	5¢	4¢	5¢	4¢	5¢	4¢
Connecticut	26 (0–26)	No Stamps										SU3 SU3 SH SI CI	AM SI E E AM AM G4	E G(3)4 E AM	AM	
		Stamps								SU3	AR	MO SU3	SH	SI(3) E SH(3) T		
Columbia Pike (west)	4 (0–4)	No Stamps									AM(6)	T G3	E			
		Stamps														
	16 (1–15)	No Stamps				HESS		E(8)	E(6) SH	G1 M G1	AM(6)	SU3				
		Stamps							SH M	SU1 R T		SI T				
	12 (0–12)	No Stamps							E(6)	AR MO		E G(3)3 AM CI				
		Stamps										SU3 SH E AR	AM			

Code: No Stamps (above dashed line) / Stamps (below dashed line)

Brands: E —Esso; AM—American; SH —Shell; T —Texaco; SU —Sunoco; SI —Sinclair; G —Gulf; CI —Citgo; MO—Mobil; AR —Arco; PB —Private Brand

—— —indicates posting price
(?) —premium differential other than indicated
G3 —3 indicates second digit of tane price (e.g. 33.9¢).

The fourth artery illustrating zone pricing is Columbia Pike. Prior to the big-volume Hess station five majors had reduced their regular price by 2.0¢ per gallon and approximately one-half of the major-brand stations posted their prices. In addition, three of the four tane prices had been reduced to 1.0¢ above Hess's regular price. The stations in clusters before, or after, those leading up to Hess generally priced at the normal major-brand price and did not post their prices.

Notes

1. An Introduction to Gasoline Marketing and Its Problems

1. U.S., Federal Trade Commission, *Report on Anticompetitive Practices in the Marketing of Gasoline*, Mimeographed (Washington, D.C.: FTC, 1967).

2. U.S., Congress, House, Select Committee on Small Business, Subcommittee No. 4 on Distribution Problems, Federal Trade Commission, *Industry Conference on Marketing of Automotive Gasoline*, 89th Cong., 1st sess.

3. *National Petroleum News Factbook Issue* (Mid-May 1971), p. 110.

4. FTC, *Anticompetitive Practices in the Marketing of Gasoline*, pp. 39–40.

2. Marketing Style of Major Oil Companies

1. "Worst U.S. Market: Could It Happen to Yours?" *National Petroleum News* (April 1964), pp. 22–32.

2. Several of the major oil companies testified to the importance of their brand names during Senate hearings. See U.S., Congress, Senate, Committee of the Judiciary, Subcommittee on Antitrust and Monopoly, *Marketing Practices in the Gasoline Industry*, Part 2, 91st Cong., 2nd sess., July 19–22, 1971.

3. Jeremy Main, "Meanwhile Back at the Gas Pump—A Battle for Markets," *Fortune* (June 1969), pp. 108–09.

4. "Not So Dumb," *Forbes* (July 15, 1969), p. 54.

5. U.S., Congress, House, Select Committee on Small Business, Subcommittee No. 4 on Distribution Problems, Federal Trade Commission, *Industry Conference on Marketing of Automotive Gasoline*, Vol. 1, 89th Cong., 1st sess., p. 711.

6. *Marketing Practices in the Gasoline Industry*, Part 2.

7. *Oil Week* (January 22, 1968), p. 6.

8. Statement of D. W. Calvert, executive vice president of Williams Brothers Pipeline, *Marketing Practices in the Gasoline Industry*, Part 2.

9. Warren C. Platt, "40 Great Years—the Story of Oil's Competition," *National Petroleum News* (March 9, 1949), p. 42.

10. John G. McLean and Robert William Haigh, *The Growth of the Integrated Oil Companies* (Boston: Graduate School of Business Administration, Harvard University, 1954), pp. 270–71.

11. Standard Oil of California v. United States, 337 U.S. 293 (1949).

12. Richfield Oil Corporation v. United States, 343 U.S. 922 (1952).

13. "Fair Trade in Jersey: Is Trouble Ahead?" *National Petroleum News* (September 1958), p. 122.

14. *National Petroleum News* (May 1970), p. 48.

15. Melvin G. de Chazeau and Alfred E. Kahn, *Integration and Competition in the Petroleum Industry* (New Haven: Yale University Press, 1959), p. 367.

16. Garrett Hardin, "The Tragedy of the Commons" in *The Environmental Handbook* (New York: Ballatine Books, Inc., 1970), pp. 36–37.

17. "New Humble's No. 1 Marketer Looks at Oil Marketing Today," *National Petroleum News* (February 1960), p. 92.

18. "They're Holding Feet to the Fire at Jersey Standard," *Fortune* (July 1970), p. 130.

19. "Conoco's Sigler Raps Fixation on Volume and Overbuilding," *Gasoline Retailer* (March 18, 1970), pp. 2 and 19.

20. "Riding the Credit Boom: New Ideas Are Coming Fast," *National Petroleum News* (December 1957), p. 94.

21. "Economist Takes a Look at Oil Credit Card Profitability," *The Oil Daily* (June 24, 1969), p. 8.

22. "Mailing Unsolicited Cards Help Boost Gallonage, Shell Oil Company Says," *Gasoline Retailer* (October 1, 1969), p. 6.

23. "Credit Card Merchandising: Can It Stop the Cost Spiral?" *National Petroleum News* (October 24, 1964), p. 101.

24. "Mobil Uses the Phone to Reduce Card Fraud," *Gasoline Retailer* (October 16, 1968), p. 36.

25. "Credit-Card Bonanza—and the Cost of Exploiting It," *National Petroleum News* (October 1964), p. 101.

26. "Bank Cards Continue March into Oil Territory," *National Petroleum News* (November 1969), p. 81.

27. "Oil Companies Puzzle: How to Get Millions to Use Their Cards," *National Petroleum News* (June 1967), p. 92.

28. "Games Mean Gallonage, But for How Long?" *National Petroleum News* (July 1967), p. 66.

29. Federal Trade Commission Hearings, *Trade Regulation Rules Regarding Games of Chance in the Food Retailing and Gasoline Industries*, Hearing Transcript, April 11, 1969, p. 658.

30. Figures compiled and released by Small Business Subcommittee, U.S. House of Representatives during their investigation of games of chance—Summer 1968.

31. "Shell Oil Company and What's Ahead for the 70's," *National Petroleum News* (July 1969), p. 75.

32. "The Battle to Build Stations," *National Petroleum News* (September 1969), pp. 75–76.

33. "Big Problems in '70's for Big 'A,' " *National Petroleum News* (December 1970), p. 62.

34. McLean and Haigh, p. 290.

35. McLean and Haigh, p. 290.

36. U.S., Congress, House, Subcommittee No. 4 on Distribution Problems, Select Committee on Small Business, Statement of Sidney Zagri, *Impact Upon Small Business of Dual Distribution and Related Vertical Integration*, Vol. 3, 89th Cong., 1st sess., p. 722.

37. "NCPR Executive Blasts Lease Policies," *Gasoline Retailer* (June 17, 1970), p. 11.

38. Testimony of Sidney Goldin, *Impact Upon Small Business of Dual Distribution and Related Integration*, pp. 728–29.

39. U.S. Representative to the Congress, John D. Dingell, Federal Trade Commission Hearings, *Trade Regulation Rules Regarding Games of Chance in the Food Retailing and Gasoline Industries*, Hearing Transcript, February 24, 1963, p. 13.

40. Testimony of W. R. Pierson, vice president of marketing, American Oil Company, *Marketing Practices in the Gasoline Industry*, Part 1, pp. 165–66.

41. Stephen Dobyns, "Standard Dealers Balk at Plan to Raise Prices," *The Detroit News* (April 1, 1971), p. 6-c and "Price-cutters Charge Purge by Mobil," *The Detroit News* (April 2, 1971), p. 13-a.

42. "Fair Trading: What Is Its Future," *National Petroleum News* (December 1956), pp. 90–91.

43. FTC, *Industry Conference on the Marketing of Automotive Gasoline*, p. 429.

44. Harvey L. Vredenburg, *Trading Stamps in Service Stations* (Iowa City, Iowa: Bureau of Business Research, 1959), pp. 19 and 22.

45. Federal Trade Commission, *Trade Regulation Rule for Games of Chance in the Food Retailing and Gasoline Industry*, Mimeographed (Washington, D.C.: FTC, 1969), pp. 22–23; and *Impact on Small Business of Dual Distribution Hearings and Related Vertical Integration*, Vol. 3, *op. cit.*, pp. 511 and 677–78.

46. "Gasoline Dealers: Fed Up and Fighting Mad," *Sales Management* (February 1, 1970), p. 27.

47. "Dealer Tells Why He Sued Texaco in Maine Termination Controversy," *Gasoline Retailer* (May 20, 1970), pp. 3, 14, and 15.

48. U.S., Federal Trade Commission, *Report on Anticompetitive Practices in the Marketing of Gasoline*, Mimeographed (Washington, D.C.: FTC, 1967), p. 42.

49. "Service Station Men Say Big Oil Companies Keep Them in Bondage," *Wall Street Journal* (October 7, 1969), p. 1.

50. "NCPR Executive Blasts Lease Policies," *Gasoline Retailer* (June 17, 1970), pp. 1 and 8.

51. "Dealer Goals Endorsed by Senator and Rep.," *Gasoline Retailer* (May 20, 1970), p. 1.

52. FTC, *Industry Conference on the Marketing of Automotive Gasoline*, p. 32.

53. "How Phillips Marched into Dixie," *National Petroleum News* (November 1956), pp. 104–05.

54. "How Phillips Is Growing in Ohio," *National Petroleum News* (March 1961), p. 91.

55. "Phillips Enters East," *National Petroleum News* (April 1961), p. 106.

56. FTC, *Industry Conference on the Marketing of Automotive Gasoline*, p. 31.

57. FTC, *Industry Conference on the Marketing of Automotive Gasoline*, p. 31.

58. For a somewhat different and more detailed explanation of the cost associated with selling major-brand gasoline see the statement of Joseph D. Harnett, vice president of marketing, Standard Oil of Ohio in FTC, *Industry Conference on Marketing of Automotive Gasoline*, p. 1921.

3. Marketing Style of Price Discounters

1. See discussion of history of trackside operators by Robert L. Bryson Jr., *The Evolution of Private Brand Marketers in the Petroleum Industry*, D.B.A. dissertation, Indiana University, 1965, pp. 32–43.

2. "Money to Be Made: The Oil Marketing Story," *National Petroleum News* (February 1969), p. 118.

3. Paul Blazer, "Reduction of Stations Necessary to Solve High Gasoline Marketing Cost," *National Petroleum News* (February 2, 1936), p. 38.

4. Edmund C. Learned, "Pricing of Gasoline: A Case Study," *Harvard Business Review*, Vol. 26 (1948), p. 727.

5. "Self-Service Stations Use Glamour, 5¢ Lower Price to Attract Business," *National Petroleum News* (January 9, 1948), pp. 33–34.

6. "Self-Service Stations Still Thriving on West Coast—Could They Succeed in Other Areas?," *National Petroleum News* (November 3, 1948), pp. 26–27.

7. "Self-Service Stations Use Glamour," pp. 33–34.

8. "Self-Service Stations Still Thriving on West Coast," pp. 40–41.

9. "Self-Service Stations Use Glamour," pp. 33–34.

10. Don Sweeney, "California's Self-Service Stations Still in Limelight," *National Petroleum News* (May 26, 1948), p. 9.

11. "Self-Service Stations Still Thriving on West Coast," pp. 28–29.

12. "Self-Service Stations Still Thriving on West Coast," p. 30.

13. Paul Wollstadt, "Multi-Pumps: What Makes Them Tick," *National Petroleum News* (October 25, 1950), p. 20.

14. Paul Wollstadt, "Most Major Companies Planning Multi-Pumps: Reduction in Operating Cost Spurs Action," *National Petroleum News* (October 18, 1950), pp. 19–21.

15. "Conoco Marketer Advises Iowa Jobbers: Price Props to Be Around Awhile," *The Oil Daily* (February 18, 1971), pp. 1 and 4.

16. Operating survey for 1966 compiled by Society of Independent Gasoline Marketers with headquarters in Clayton, Missouri.

17. "Merchandise Keeps Independent Growing," *National Petroleum News* (March 1969), pp. 108–09.

18. "Can a Private Brander Swing with the Times," *National Petroleum News* (August 1969), p. 72.

19. "Can a Private Brander Swing with the Times," p. 72.

20. "Can a Private Brander Swing with the Times," pp. 73–74.

21. "An Independent Refiner Builds a Brand," *National Petroleum News* (June 1956), pp. 100–02.

22. Bob Latimore, "Self-Service," *Vend* (August 15, 1962), pp. 32–33.

23. Latimore, p. 35.

24. Latimore, p. 34.

25. "That 'Gasamat' Man Rides High in the West," *National Petroleum News* (January 1969), p. 54.

26. Latimore, pp. 33–34.

27. "That 'Gasamat' Man Rides High in the West," p. 50.

28. "New Look at Self-Service," *National Petroleum News* (April 1965), pp. 106–07.

29. "Self-Service Gasoline: The Phenomenon of the 1960's," *National Petroleum News* (June 1969), pp. 74–80.

30. "A New Self-Service Market Takes Root," *National Petroleum News* (May 1969), p. 102.

31. "A New Self-Service Market Takes Root," pp. 100–01.

32. "Self-Service Picks Up Steam, Many Climb Aboard," *National Petroleum News* (June 1971), pp. 62–63.

33. "Self-Service Picks Up Steam," pp. 62–63.

4. Competition. Conflict, and Exclusion

1. Jeremy Main, "Meanwhile Back at the Gas Pump—A Battle for Markets," *Fortune* (June 1969), p. 108.

2. Joseph Palamountain, *The Politics of Distribution* (Boston: Harvard University Press, 1955), p. 29.

3. U.S., Congress, Senate, Committee of the Judiciary, Subcommittee on Antitrust and Monopoly, Testimony of W. R. Pierson, vice president of marketing, American Oil Company, *Marketing Practices in the Gasoline Industry*, Part 1, 91st Cong., 2nd sess., p. 152.

4. *Platt's Oil Price Handbook and Oilmac: 1968 Prices* (New York: McGraw-Hill Book Company, 1969), p. 190.

5. *A Factual Guide to Oil Industry Issues* (New York: American Petroleum Institute, 1969), p. 3.

6. Pierson, *Marketing Practices in the Gasoline Industry*, p. 155.

7. U.S., Congress, House, Select Committee on Small Business, Subcommittee No. 4 on Distribution Problems, Federal Trade Commission, Statement of H. J. Peckheiser, executive vice president, Mobil Oil Company, FTC, *Industry Conference on Marketing of Automotive Gasoline*, Vol. 1, 89th Cong., 1st sess., p. 422.

8. Statement of E. R. Bradley, general sales manager, Sun Oil Company, FTC, *Industry Conference on Marketing of Automotive Gasoline*, p. 434.

9. "Unbranded Gasoline Shakes Pricing Structure," *National Petroleum News* (October 29, 1952), p. 17.

10. *National Petroleum News* (November 1965), p. 83.

11. "How Private Brands are Burgeoning in the Interior West," *National Petroleum News* (October 1963), p. 24.

12. Testimony of Frank H. Staub, vice president of East Coast marketing, Shell Oil Company, FTC, *Industry Conference on Marketing of Automotive Gasoline*, p. 1791.

13. "Worst U.S. Market: Could It Happen to Yours?" *National Petroleum News* (April 1964), p. 24.

14. Frank H. Staub, FTC, *Industry Conference on Marketing of Automotive Gasoline*, pp. 1781–85.

15. *National Petroleum News Bulletin* (September 5, 1961), p. 4.

16. Richard M. Clewett, Ralph Westfall, Harper Boyd, *Cases in Marketing Strategy* (Homewood, Illinois: Richard D. Irwin, 1964), p. 233.

5. THE ONE-CENT DIFFERENTIAL WARS

1. Ralph Cassidy, Jr., "Price Warfare in Business Competition: A Study of Abnormal Competitive Behavior," Occasional Paper No. 1 (East Lansing: Bureau of Research, Michigan State University, 1963).

2. U.S., Congress, Senate, Committee of the Judiciary, Subcommittee on Antitrust and Monopoly, Testimony of Fred C. Allvine, *Marketing Practices in the Gasoline Industry*, Part 1, 91st Cong., 2nd sess., p. 90.

3. "Don't Tread on Me—Anymore," *National Petroleum News* (May 1958), p. 82.

4. "Gasoline Wars—and How to Stop Them," *National Petroleum News* (December 16, 1953), p. 33.

5. "Don't Tread on Me—Anymore," p. 82.

6. "Don't Tread on Me—Anymore," p. 82.

7. "Normal Prices? What Normal Prices, Ask Southwest Marketers?" *National Petroleum News* (October 1961), p. 28.

8. "Wanted Answers to These Questions," *National Petroleum News* (June 1961), p. 81.

9. "Don't Tread on Me—Anymore," p. 82.

10. U.S., Congress, House, Select Committee on Small Business, Subcommittee No. 4 on Distribution Problems, Federal Trade Commission, *Indus-*

try Conference on Marketing of Automotive Gasoline, Vol. 1, 89th Cong., 1st sess., pp. 683–84.

11. "Wanted Answers to These Questions," p. 81; and "More New Price Thinking on the Way," *National Petroleum News* (May 1961), pp. 72–73.

12. "What's Behind Shell's 1¢ Policy," *National Petroleum News* (August 1962), p. 82.

13. FTC, *Industry Conference on the Marketing of Automotive Gasoline*, pp. 384–85.

14. *National Petroleum News Bulletin* (October 12, 1964), p. 1.

15. "Oil," *Forbes* (January 1, 1965), p. 27.

16. "Latest in Mergers: Atlantic-Richfield Deal," *National Petroleum News* (October 1965), p. 75.

17. "After Full Year of One Cent Differential: West Coast Independents Restless," *National Petroleum News Bulletin* (December 31, 1962), p. 1.

18. "Far West Independents Gaining Ground While Majors Plot a Retaliation," *National Petroleum News* (October 1969), p. 51.

19. "West Coast: The Rush Is On," *National Petroleum News* (August 1960), p. 85.

20. "How Independents Feel About 1¢ Differential," *National Petroleum News* (November 1962), p. 37; and "What's Ahead for the Secondary Brands of the Major Oil Companies?" *National Petroleum News* (July 1962), p. 20.

21. "More New Price Thinking on the Way? It's a Must Says Two Independents," *National Petroleum News* (May 1961), p. 73.

22. FTC, *Industry Conference on the Marketing of Automotive Gasoline*, pp. 647 and 648.

23. FTC, *Industry Conference on the Marketing of Automotive Gasoline*, p. 122.

24. FTC, *Industry Conference on the Marketing of Automotive Gasoline*, p. 692.

25. FTC, *Industry Conference on the Marketing of Automotive Gasoline*, pp. 48 and 49.

26. *National Petroleum News Bulletin* (August 13, 1962), p. 1.

6. Subregular-grade Gasoline Price Wars

1. "Will Sun's Five Grades Start a Trend?" *National Petroleum News* (April 1956), pp. 106–07.

2. "The Chips Are Down in the Big Fight for Premium Sales," *National Petroleum News* (April 1957), pp. 94–96.

3. "Sub-Regular Blending," *National Petroleum News* (June 1961), p. 86.

4. "Sub-Regular Blending," p. 86.

5. Don G. Campbell, "Compact Behind War," *The Indianapolis Star* (September 8, 1961), p. 30.

6. "What's in a Name," *Forbes* (December 15, 1961), pp. 14–15.

7. "Market Trends," *National Petroleum News Bulletin* (September 5, 1961), p. 4.

8. Records of Private-Brand Gasoline Association of Indiana.

9. "Big Price Test: Can Marketers Weather a Year of Chaos?" *National Petroleum News* (January 1962), p. 93.

10. Letter from senior vice president of Gulf Oil Company to Independent Oil Marketers' Association of Indiana dated August 14, 1961.

11. Advertisement in *The Indianapolis Star* (January 23, 1962), p. 15.

12. Draft of a letter by Louis E. Kincannon, president of Golden Imperial Oil Company, Indianapolis, Indiana.

13. "How Gulf Will Test and Evaluate Its 'Combat' Gasoline, Gulftane," *National Petroleum News* (September 11, 1961), p. 1.

14. "Big Price Test: Can Marketers Weather a Year of Chaos?" *National Petroleum News* (January 1962), p. 94.

15. "First-Half Profits Up 3.4%; Last Half Looks Dimmer," *National Petroleum News Bulletin* (August 13, 1962), p. 1.

16. "Summer Brings Gasoline—Price Disintegration," *National Petroleum News Bulletin* (July 16, 1962), p. 1.

17. "Gasoline Price Structures Under Increasing Pressure," *National Petroleum News Bulletin* (September 17, 1962), p. 1.

18. "Price Protection Pot Boils Over," *National Petroleum News* (October 1962), p. 87; and "Price Supports Come Off, Leaving Big Question," *National Petroleum News Bulletin* (September 4, 1962), p. 1.

19. "New Wrinkles Showing Up in Gasoline Markets," *National Petroleum News Bulletin* (October 8, 1962), p. 1.

20. "Gasoline Prices Moving Downward As Piecemeal Cuts Erode Markets," *National Petroleum News Bulletin* (May 6, 1963), p. 1.

21. "Gulf Embraces 1¢ Spread for Gulftane," *National Petroleum News Bulletin* (August 26, 1963), pp. 2 and 3.

22. "Price Changes Are in the Air—But Is Price Improvement?" *National Petroleum News Bulletin* (March 23, 1964), p. 2.

23. "Worst Market's Getting Better? Detroit Price Levels Move Up," *National Petroleum News Bulletin* (July 20, 1964), p. 1.

24. Halsey Peckworth, "Cost Price Squeeze Puts Independents in Double Bind," *National Petroleum News Bulletin* (September 8, 1964), pp. 1 and 2.

25. Halsey Peckworth, "What's Going to Happen to the Independents?" *National Petroleum News Bulletin* (November 2, 1964), p. 1.

26. "Price Advance," *National Petroleum News* (May 1965), p. 59.

27. *National Petroleum News Bulletin* (February 6, 1967), p. 4.

28. This conclusion is supported by the testimony of Alfred E. Kahn before the Hart Committee. U.S., Congress, Senate, Committee of the Judiciary, Alfred E. Kahn, "Mergers in the Petroleum Industry and Problems of the Independent Refiner," *Economic Concentration*, Part II, 91st Cong., 1st sess., pp. 562–609.

29. "Gasoline Markets Worsen; Some Nine-Month Profit Improvement," *National Petroleum News Bulletin* (October 27, 1962), p. 1.

30. "Behind Those Bright Earnings Reports and Swelling Volumes . . . ," *National Petroleum News Bulletin* (August 3, 1964), pp. 1 and 4.

31. "Oil," *Forbes, 17th Annual Report on American Industry* (January 1, 1965), p. 28.

32. *National Petroleum News Bulletin* (February 8, 1965), p. 1.

33. U.S., Congress, House, Select Committee on Small Business, Subcommittee No. 4 on Distribution Problems, Testimony of G. C. Briggs, general sales manager, Standard Oil of California, *Impact on Small Business of Dual Distribution and Related Vertical Integration*, Vol. 3, 89th Cong., 1st sess., p. 719.

34. "Energy," *Forbes, 21st Annual Report on Industry* (January 1, 1969), p. 113.

35. *Impact on Small Business of Dual Distribution and Related Vertical Integration*, Vol. 3, p. 689.

36. Halsey Peckworth, "Cost Price Squeeze Puts Industry in Double Bind," *National Petroleum News Bulletin* (September 8, 1964), pp. 1 and 4.

37. Halsey Peckworth, "What's Going to Happen to the Independents?" *National Petroleum News Bulletin* (November 2, 1964), p. 1.

38. "Hess Will Buy Delhi Units in $25 Million Cash Deal," *National Petroleum News Bulletin* (January 28, 1963), p. 3.

39. "Fina Buys Cosden Petroleum, Station Will Be Fina Brand," *National Petroleum News Bulletin* (March 4, 1963), p. 1.

40. " 'Why Sell Gasoline,' Ask West Coast Independent Marketers, 'If I Can Make More Selling Stations,' " *National Petroleum News* (November 1961), p. 17.

41. Mark Edmond, "What's Happening in World's Biggest Gasoline Market," *National Petroleum News* (March 1969), p. 27; and interviews with West Coast suppliers and marketers.

42. "Big Price Test: Can Marketers Weather a Year of Chaos?" *National Petroleum News* (January 1962), p. 93.

43. Marvin Reid, "Spot Reports: As Gulftane Moves Into New Markets," *National Petroleum News* (January 1962), pp. 97 and 98.

44. "Texas Jobber 'King' Is a Mystery Man to Many," *National Petroleum News* (May 1970), p. 83.

45. "In a Changing Market Place, Must Private Branders Change?" *National Petroleum News* (May 1968), p. 72.

7. Price Protection Programs

1. Office of Emergency Preparedness, *Report on Crude Oil and Gasoline Price Increases of November 1970* (April 1971).

2. "Senate Gasoline Probe Expands," *U.S. Oilweek* (June 29, 1970), p. 1.

3. *Platt's Oilgram Price Service* (April 21, 1970), p. 2.

4. Stephen Dobyns, "Price-cutters Charge Purge by Mobil," *The Detroit News* (April 2, 1971), p. 13-a; and "Gas War: Standard Dealers Balk at Plans to Raise Prices," *The Detroit News* (April 1, 1971), p. 6-c.

5. U.S., Congress, House, Select Committee on Small Business, Subcommittee No. 4 on Distribution Problems, Federal Trade Commission, *Industry Conference on the Marketing of Automotive Gasoline*, Vol. 1, 89th Cong., 1st sess., p. 95.

6. *Platt's Oilgram Price Service* (February 9, 1970), p. 2; (June 18, 1970), p. 2.

7. "Behind the Pricing Plan that Jolted Midwest Markets," *National Petroleum News* (March 1961), p. 95.

8. "Suppliers Move to Stop Cross-Hauls, Subsidy Abuse," *National Petroleum News Bulletin* (December 24, 1962), p. 4.

9. U.S., Congress, Senate, Committee of the Judiciary, Subcommittee on Antitrust and Monopoly, Testimony of Fred C. Allvine, *Marketing Practices in the Automotive Gasoline Industry*, Part 1, 91st Cong., 2nd sess., p. 90.

10. "Conoco Executive Raps Granting of Allowances During 'Gas' Wars," *Oil Daily* (October 13, 1970), pp. 1 and 16.

8. Buy-outs of Price Marketers

1. Harper Boyd and Ralph Westfall, "Oklahoma Oil Company," in *Cases in Marketing Management* (Homewood, Illinois: Richard D. Irwin, 1961), pp. 264–67.

2. "Gasteria's Dream: A Franchised Trademark for Private Branders," *National Petroleum News* (November 1957), p. 125.

3. "Oklahoma Converts Gasteria," *National Petroleum News* (May 1958), p. 78.

4. Frank Breese, "West Coast: The Rush Is On!" *National Petroleum News* (August 1960), p. 85.

5. "Wilshire's All Dressed Up and Ready to Go for the Top," *National Petroleum News* (September 1957), p. 115; and "What Gulf is Doing to 'Go Nationwide' in Marketing," *National Petroleum News* (October 1965), p. 76.

6. "Next Big Merger: Union Oil and Atlantic Refining?" *National Petroleum News* (January 1964), pp. 61 and 107.

7. "How Spur Hit the Jackpot," *National Petroleum News* (May 1958), p. 8; and "Murphy Expands," *National Petroleum News* (October 1960), p. 69.

8. "Ingram: 'We'll Build a Brand in Five Years,' " *National Petroleum News* (February 1958), p. 133.

9. Arthur M. Louis, "Leon Hess Never Plays It Safe," *Fortune* (January 1970), p. 141.

10. "Signal Oil Quitting Marketing," *National Petroleum News* (October 1970), p. 49.

9. Vertical Integration and Monopoly Power

1. Melvin G. de Chazeau and Alfred E. Kahn, *Integration and Competition in the Petroleum Industry* (New Haven: Yale University Press, 1959), p. 115.
2. de Chazeau and Kahn, p. 117.
3. George S. Wolbert, Jr., *American Pipe Lines* (Norman: University of Oklahoma Press, 1951), p. 9.
4. Henrietta M. Larson and Kenneth W. Porter, *History of Humble Oil and Refining Company* (New York: Harper and Brothers, 1959), pp. 169 and 698.
5. George S. Wolbert, Jr., pp. 165–99.
6. Larson and Porter, p. 610.
7. U.S. Department of Justice, *Fourth Report of the Attorney General on the Interstate Oil Compact* (Washington, D.C.: U.S. Government Printing Office, 1959), p. 34 and U.S. Department of Interior, "Refining and Petro-chemical Inputs," April 5, 1971.
8. "How a Major Shifts from 'Defense' to 'Offense,'" *National Petroleum News* (March 1965), pp. 100–01.
9. Office of Emergency Preparedness, *Report on Crude Oil and Gasoline Price Increases of November 1970* (Washington, D.C., April 1970), p. 40.
10. "The Luck of the Drill Bit," *Forbes* (January 15, 1970), p. 16.
11. *Forbes* (June 1, 1963), p. 3.
12. "Energy," *Forbes* (January 1, 1969), p. 108.
13. "Inside Report on a $370,000,000 Pipeline," *National Petroleum News* (February 1965), p. 66.
14. 42 USC 1857 Section 202 (b) (1) (A).

10. Preferential Treatment of Crude Oil

1. Henrietta M. Larson and Kenneth W. Porter, *History of Humble Oil and Refining Company* (New York: Harper and Brothers, 1959), pp. 302–06.
2. U.S., Congress, Senate, Committee of the Judiciary, Subcommittee on Antitrust and Monopoly, Testimony of economist Walter J. Mead, *Government Intervention in the Market Mechanism*, Part 1, 91st Cong., 1st sess., p. 80.
3. U.S., Department of Justice, *First Report of the Attorney General on the Interstate Compact to Conserve Oil and Gas* (Washington, D.C.: U.S. Government Printing Office, 1956), p. 59.

4. Walter J. Mead, *Government Intervention in the Market Mechanism,* p. 83.

5. *Interstate Compact to Conserve Oil and Gas,* p. 39.

6. *First Report . . . Interstate Compact to Conserve Oil and Gas,* p. 46.

7. U.S., Department of Justice, *Second Report of the Attorney General on the Interstate Compact to Conserve Oil and Gas* (Washington, D.C.: U.S. Government Printing Office, 1957), pp. 39–48.

8. "Why the Oil Giants Are Under the Gun," *Business Week* (October 25, 1969), pp. 82–83.

9. Testimony of economist Paul H. Frankell, *Government Intervention in the Market Mechanism,* p. 198.

10. "Four Majors Schedule Import Cuts," *National Petroleum News* (June 9, 1954), p. 18.

11. Glenn M. Green, Jr., "The Import Controversy: Its Pros and Cons," *National Petroleum News* (March 18, 1953), p. 30.

12. Allan T. Demaree, "Our Crazy Costly Life with Oil Quotas," *Fortune* (June 1969), p. 107.

13. Glenn M. Green, Jr., p. 29.

14. Allan T. Demaree, p. 107.

15. Testimony of economist M. A. Adelman, *Government Intervention in the Market Mechanism,* pp. 8–12.

16. Testimony of George P. Shultz, *Government Intervention in the Market Mechanism,* pp. 1707–09.

17. Jim Drummond, "Haider Encouraged After Talk With President," *Oil Daily* (November 12, 1969), p. 1.

18. See *Oil Daily* of December 1, 8, and 11.

19. Jim Collins, "Long-Awaited Task Force Report Released: Nixon Decides Against Immediate Change," *Oil Daily* (February 24 ,1970), p. 1.

20. "Nixon Import Move: Big Oil Is Smiling," *National Petroleum News* (October 1970), p. 52.

21. "Should the Depletion Allowance Be Cut," *National Petroleum News* (May 1959), p. 91.

22. Ronnie Dugger, "Oil and Politics," *Atlantic Monthly* (September 1959), pp. 74–75; and Robert Engler, *The Politics of Oil* (New York: Macmillan, 1961), Chapters 12 and 13, pp. 341–94.

23. "Is Depletion Allowance a 'Must?' " *National Petroleum News* (August 13, 1952), pp. 34–35; Testimony of Harold McClure, president of the Independent Petroleum Association of America, *Government Intervention in the Market Mechanism,* p. 619; and "Should the Depletion Allowance Be Cut?" pp. 34–35.

24. Harold M. McClure, Jr., "Nationwide Shock-waves Would Follow an Increase in Oil's Tax Burden," *Independent Petroleum Monthly* (April 1969), pp. 16–23.

25. Ronnie Dugger, p. 87.

26. Clark W. Breeding and A. Gordon Burton, *Taxation of Oil and Gas Income* (New York: Prentice-Hall, 1954), p. 158.

27. "Why the Oil Giants Are Under the Gun," *Business Week* (October 25, 1969), p. 88.

28. Statement of Senator Edward M. Kennedy, *Government Intervention in the Market Mechanism*, p. 15.

29. "Why the Oil Giants Are Under the Gun," pp. 82–83.

30. Testimony of Kenneth E. Hall, *Government Intervention in the Market Mechanism*, p. 808.

31. Melvin G. de Chazeau and Alfred E. Kahn, *Integration and Competition in the Petroleum Industry* (New Haven: Yale University Press, 1959), p. 451.

11. Conclusions and Recommendations

1. Bill Mullins, "Price Props to Be Around Awhile," *Oil Daily* (February 18, 1971), pp. 1 and 4. Also interviews with major oil company executives.

2. Melvin G. de Chazeau and Alfred E. Kahn, *Integration and Competition in the Petroleum Industry* (New Haven: Yale University Press, 1959), pp. 564 and 565.

Index